JUST A THEORY

JUST A THEORY

| Exploring the Nature of Science |

MOTI BEN-ARI

 Prometheus Books

59 John Glenn Drive
Amherst, New York 14228-2197

Published 2005 by Prometheus Books

Inquiries should be addressed to
Prometheus Books
59 John Glenn Drive
Amherst, New York 14228-2197
VOICE: 716-691-0133, ext. 207
FAX: 716-564-2711
WWW.PROMETHEUSBOOKS.COM

09 08 07 06 05   5 4 3 2 1

Library of Congress Cataloging-in-Publication Data

Ben-Ari, M., 1948–
    Just a theory : exploring the nature of science / Moti Ben-Ari.
        p. cm.
    Includes bibliographical references and index.
    ISBN 1-59102-285-1 (alk. paper)
    1. Science—Theory reduction. 2. Research. 3. Science—Methodology.
4. Scientism. I. Title.

Q175.B4195 2003
501—dc22
                                                            2004029953

Printed in the United States of America on acid-free paper

For my sons Hezi and Yoav

Contents

Preface

Scientific terminology and scientific claims surround us, and many of today's important political issues revolve around science. Should we change our economy to prevent global warming or has the threat been exaggerated? What should we do with nuclear waste? How should health authorities deal with infectious diseases like AIDS, tuberculosis, and malaria? Is money spent on space exploration being wasted? The debate on these and other similar questions is often surprisingly superficial, primarily because few people know how to read and interpret reports of scientific results even when they have been written by competent journalists and science writers. Fewer still can analyze public debates among scientists holding different opinions on a subject.

In order to understand scientific issues, you have to have a clear picture of what science is and what scientists do. Some people claim that the theory of evolution is "just a theory." What exactly is a scientific theory? Advertisements claim that a certain laundry liquid or headache remedy has been "scientifically proven" to be better than a competing product. What, if anything, does it mean for a claim to be scientifically proven? A research report claims that eating a certain food will "reduce the chance" of suffering a heart attack by 14 percent. How is such research performed and what is the relevance of a 14 percent chance?

The aim of this book is to provide a modern overview of the *nature of science*, so that you can understand and analyze scientific results and debates that you are exposed to every day. It will help you distinguish claims that are provisional and debatable, from claims that are so well established that rejecting them drives one over the border that divides real science from *pseudoscience*, which are activities that illegitimately wrap themselves in the mantle of science. To achieve this aim, we must examine the nature of science in depth, in order to transcend the naive image of a man in a lab coat comparing laundry detergents.

The explosion of scientific knowledge in the twentieth century was accompanied by deep thinking and extensive research on the nature of science, traditionally divided into the philosophy, history, and sociology of science.

The *philosophy of science* focuses on epistemology, the branch of philosophy that deals with the concept of knowledge. What is scientific knowledge? Few people appreciate that modern science is quite limited in scope and restricts itself to description and explanation of natural phenomena; on purpose, science does not deal with *purpose*. How is scientific knowledge obtained? Is there a scientific method that will lead inexorably to truth? If so, can this method be applied to nonscientific fields? What demarcates scientific knowledge from nonscientific knowledge? The issue of demarcation is particularly important, because of the pretensions of nonscientific activities to be scientific. Pseudosciences such as astrology are enormously popular. Though sometimes harmless, pseudosciences pose a potential danger to individuals and to society when crucial decisions are based on claims that totally lack empirical evidence.

At its core, science is ahistorical, so that if all books and Web pages were to be destroyed except for a handful of "dry" textbooks of mathematics, physics, chemistry, and biology, science would not be hopelessly impoverished in the way that culture would be impoverished if the plays of Shakespeare or the paintings of Monet were destroyed and all we were left with were textbooks describing these works. However, we do live within a historical context and a study of the *history of science* is just as much a part of our heritage as the study of political, economic, or social history. The study of science without its history might not damage our knowledge of its results, but it is likely to impair our ability to teach science and to relate to science as it affects our lives.

The third aspect of the nature of science is the *sociology of science*, the study of which is relatively recent. Science is "done" by individuals, but these individuals work in groups and have different roles: they are professors, graduate students, research scientists, laboratory technicians and engineers. They work in universities and in industrial and governmental research institutions. Sociologists and anthropologists can investigate the institutions and culture of science using the same methods and tools that they use to investigate any society and culture. Unfortunately, some of the *postmodernist* study of the social aspects of science has degenerated into antiscience, claiming that science is a purely social activity and that the "real world"—if it even exists—is irrelevant to the practice of science. Any discussion of the nature of science has to explain how to integrate the epistemological claim that science can discover the truth about the universe with the practice of science by fallible humans who are members of social institutions and cultures.

The book starts with a discussion of a widely held caricature of the process of science, the naive inductive-deductive method. Then we will extensively explore the concept of a scientific theory and explain why it is the central concept of the nature of science. We will illustrate the concept with two examples: the theory of gravitation and the theory of evolution by natural selection and explain why both have achieved the (high) status of theories. Next, we will examine the basic terminology used by scientists and show how it differs from the day-to-day use of the same terms. The excursion into the philosophy of science continues with Karl Popper's principle of falsification, which leads naturally into a discussion of pseudoscience. Then we turn to Thomas Kuhn's 1962 book *The Structure of Scientific Revolutions*, which was instrumental in initiating the study of the sociology of science. We summarize Kuhn's ideas and refute the postmodernistic denigration of science that grew out of an aberrant interpretation of Kuhn.

The book continues with a survey of the modern viewpoints concerning the relationship between science and religion. Then we discuss the relationship between the various sciences, especially the concept of reductionism; a misunderstanding of reductionism is used to justify some claims of pseudoscience and postmodernism. The next three chapters provide more detail on the nature of specific scientific topics: statistical methods such as those used in medical science, logic, and mathematics—the languages of science, and a case study of the nature of science as typified by modern theories in geology. Finally, we summarize speculation on the future of science.

Interspersed between the chapters, we present biographical vignettes of some famous scientists, in order to illustrate that science is done within a cultural and historical context, and—contrary to the stereotype—by scientists of widely varying personalities.

The technical content of the book is held to an absolute minimum so that it will be accessible to a wide audience. Occasionally, I have not been able to resist the temptation to include a bit of mathematics or technical detail. Except for Maxwell's and Schrödinger's equations—included only for their visual effect—the mathematics is at a high-school level. This technical material has been set off in sidebars that can be safely skipped.

There are excellent books accessible to the nonspecialist on each of the topics mentioned, and an annotated reading list has been provided in order to enable you to pursue these issues in depth.

The days are long gone when a lone "amateur gentleman" can make a significant contribution to science or even understand the details of modern scientific theories and results. Nevertheless, science is not a secret society based upon arcane rituals; science is a search for truths about the universe, and its basic principles are accessible to any educated person. This book will have served its purpose if will help you to keep an open mind, but not so open that your brains fall out.[1]

Acknowledgments: I would very much like to thank Michael Matthews, William McComas, Edmond Schonberg, and Peter Slezak for reading and commenting on drafts of the book. Needless to say, they will not agree with everything I have written, and the mistakes that remain are my responsibility.

Moti Ben-Ari
Rehovot, Israel
March 2005

1

Naive Induction:
What People Think Scientists Do

A screenplay

Galileo Galilei (1564–1642) is considered to be the first modern scientist. Here is an imaginary screenplay describing his activities as a scientist:

Galileo Galilei climbs up the Leaning Tower of Pisa carrying a bunch of objects: some shaped as balls, some as cylinders, and some as cubes; some large, some small, and some in between; some made of stainless steel, some of titanium, and some of plastic; some painted black, some painted white, and some chrome-plated (altogether, $3 \times 3 \times 3 \times 3 = 81$ objects, one with each possible combination of these properties). Galileo had obtained a governmental grant for a long-term research project, intended to decide once and for all if the time of the fall of an object depends on its weight or not. A careful experimenter, Galileo arranges for a robot arm to push pairs of objects off the parapet of the tower in order to ensure that the objects are dropped simultaneously and not affected by his conscious or subconscious actions. The computer-generated signal to the robot arms also starts an electronic clock at the instant the objects fall. When the objects hit sensors on the ground, signals are generated that cause the time of the fall of each object to be stored in the memory of the computer. The experiments takes years to carry out, not only because of the vast number of tests to be run (there are $81 \times 80/2 = 3,240$ different pairs of objects), but also because Galileo had to cancel his experiments in inclement weather in order to ensure that neither rain nor wind would compromise the accuracy of his results.

Following years of painstaking measurements, Galileo creates a database and writes a computer program to analyze the mountain of data that he has collected. A couple of years later, after the program has been debugged, graphs and charts spew forth onto Galileo's desk. Galileo spends months examining them and then calls a press conference to announce to the world: all objects dropped from a given height hit the ground at the same time. Several years later he is granted the Nobel Prize for Physics, and after accumulating a comfortable nest egg earned by giving lectures and making TV appearances, he withdraws to a comfortable retirement at his villa on the shores of beautiful Lake Como in northern Italy. *(End of screenplay).*

How many anachronisms are there in the above screenplay? I am sure that you picked up that there were no computers, no titanium balls, no robots, and no Nobel Prizes in the seventeenth century. You may also know that Galileo was tried by the Inquisition and spent his final years under house arrest near Florence, not strolling in gardens, admiring the scenery at Lake Como. But in fact, the entire story is a literary exercise: There is no evidence that Galileo ever dropped anything off the Leaning Tower of Pisa, certainly not in a public demonstration intended to show that other scientists were mistaken. In his *Dialogues Concerning Two New Sciences*, Galileo admits to having performed experiments that showed differences between the time of fall of heavy and light objects, but he clearly understood that the differences were negligible, and due to confounding factors like air resistance. More importantly, however, Galileo did not base his claim about falling bodies directly upon experimental evidence. Although he did perform experiments rolling balls down ramps, he only used the results to guide his theoretical thinking and mathematical demonstrations. His books contain no large sets of data convincingly presented in spreadsheets and graphs; in fact, they contain no data whatsoever.

Galileo, the first modern scientist

Galileo was responsible for two of the most important conceptual advances that led to the development of modern science. He was the first person to formulate universal mathematical laws to describe the motion of objects. From the time of the Greeks, mathematics had been used in astronomy: planets were assumed to move in circles around the Earth, and the problem was to find an arrangement of circles that would fit the data. Later, Nicolaus Copernicus (1473–1543) proposed that all planets, including the Earth, rotate around the Sun, and Johannes Kepler (1571–1630) found that ellipses are superior to circles for describing planetary motion. But only from the time of Galileo did scientists begin to formulate mathematical laws that were supposed to be *universally valid* so that they could used to make accurate predictions. By establishing that objects fall with constant acceleration, Galileo was not just describing the motion of a certain ball dropped off of a certain tower; instead he enabled scientists to compute the motion of any object dropped from anywhere.

The second conceptual advance attributed to Galileo is the introduction of idealizations in order to make scientific problems tractable. In the seventeenth century, it was totally impossible to perform very precise and repeatable experiments on falling objects. It was not just that they didn't have computer-controlled robots and sensors; they did not even have clocks that could measure time with reasonable accuracy and precision. The best one could do was to allow water to drip out of a container through a small orifice and then weigh the water. Galileo approached the problem of analyzing motion by performing experiments with inclined planes, as motion on a plane was slower and easier to measure than the fall of an object. Then he imagined what motion would be like if the inclined plane were frictionless and showed mathematically that motion on an inclined plane was equivalent to the motion of a falling object under the influence of gravity.

There are two idealizations here. The first is to describe motion in a nonexistent situation and then to claim that the description can be usefully applied to a real situation. Galileo described the motion of objects falling in a vacuum or rolling down frictionless ramps, and then applied the results to real situations by assuming that air resistance and friction are negligible, or alternatively, that they can be accounted for by appropriate corrections. The second idealization is to use the abstract language of mathematics for analyzing the situation, and then to claim that the equations that result could be used in real situations.

Galileo's great achievement was to identify the correct idealizations: acceleration as the rate of change of velocity per unit of time, gravitation as exerting a constant acceleration per unit of mass, and the concept of inertia, the tendency of a body to continue in motion unless acted upon by a force.

When Galileo presented his results, he rarely referred to experimental results and never provided appendices full of tables and graphs. Instead, he used thought experiments and mathematical arguments to convince the reader of the correctness of his claims and of their applicability. Here is an adaptation of his thought experiment that showed that the time of the fall of an object under the influence of gravity is independent of the mass of the object.

Let us imagine that two objects, one light and one heavy, are simultaneously dropped from a tower. How can we show that they will both arrive

Galileo's Thought Experiment

Suppose that heavy objects fall faster than light ones, so that a 20 kilogram object hits the ground 15 seconds after being dropped from a tower, while a 10 kilogram object takes 30 seconds. Consider a 20 kg object constructed of two 9.9 kg balls and a thin, but strong, bar weighing 0.2 kg (9.9 + 9.9 + 0.2 = 20). The center of the bar is fitted with a separation device, which can break the object into two pieces without significantly changing its downward motion. Let us activate the device just after the 20 kg object is released, turning it into two objects of 10 kg each. How can each ball suddenly "know" that it is no longer connected to its twin and thus decelerate so as to arrive at the ground after 30 instead of 15 seconds?

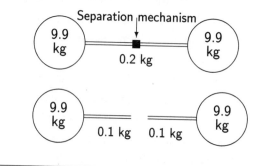

at the ground at the same time? Let us suppose to the contrary that heavier objects fall faster than light objects and show that this leads to an absurd situation. Imagine now that halfway through its fall, the heavy object breaks apart into two pieces of equal size. Well, light objects are assumed to fall slower, so somehow the two pieces "understand" that they are no longer part of a heavy object and stomp on the brakes! That is totally absurd because an object can't slow down without force being applied and there is no reason to assume that a part of an object "knows" that it belongs a larger object.

Once you know the principle, it is obvious that you have seen the effect before: you throw a snowball or clod of dirt into the air and even if it breaks apart, all the pieces reach the ground more or less together. Conversely, if two snowballs suddenly meet in midair like the rendezvous of two spacecraft, it is inconceivable that the speed at which they fall should increase, merely because some flakes stick together with no force applied in the vertical direction.

The scientific method

If you ask most people to define *the* scientific method, you would most likely receive a description of something akin to the *naive inductive-deductive method*:

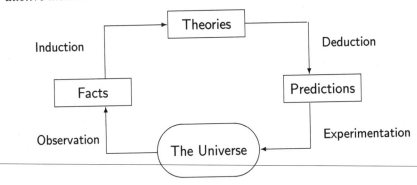

Science proceeds in a well-structured manner consisting of four successive and distinct stages. First, the scientist observes the universe, recording her observations as facts. Then, she performs a process called *induction*: by examining the facts, she develops a generalization called a *theory* that can account for these facts. Next, in the second, *deductive* stage, the scientist uses logic to predict consequences of the theory and then she carries out experiments to see if these predictions do in fact occur. If the experiments are successful, the theory has been confirmed; if not, the theory has been shown to be incorrect. Let us examine the elements of this portrait one by one.

Observation

I have portrayed observation as unstructured. The scientist walks along, examining rocks and plants or gazing at the stars, notebook in hand, writing down his observations. Or perhaps, he scoops up a sample of water to examine under the microscope back in his laboratory. But this makes no sense at all. There are too many things to observe: people, plants, animals, rocks, stars, rivers, the wind. Inevitably, a scientist's observations are focused on a topic of interest. Furthermore, the observation is not passive; the scientist will actively seek out interesting phenomena. But how does he decide what to focus on? How does he decide what is scientifically interesting? Is it simply a matter of personal preference?

When Charles Darwin (1809–1882) was young, he was a passionate collector of beetles. He was quite a competent observer, learning how to identify and classify them. But this made him a beetle collector, *not* a scientist. Observations in science must be made within a framework if they are to have any significance. Darwin spent five years as a naturalist on the ship HMS *Beagle* as it traveled around the world. He performed observations in geology and biology, gathering and preserving samples, and sketching and describing the terrain that he saw. In contrast to his beetle-collecting days, Darwin's geological observations were conducted within a framework called *uniformitarianism* that had been proposed by Charles Lyell (1797–1875). According to uniformitarianism, the geological formations that we observe were formed by gradual processes that continue to this day. Any geological observation that Darwin made could be interpreted by him as evidence in favor or against Lyell's theory.

But in biology there was as yet no theory that could help Darwin focus his observations. Plants and animals do not lie around saying, "Observe me: I can be used in the inductive stage of the formulation of the theory of evolution!" To his eternal regret, when Darwin collected the famous finches from the Galapagos Islands that would be critical in his development of the theory of evolution, he did not think to label them with the name of the specific island from which they came. At that time, naturalists saw their task as one of collecting and identifying organisms, and of comparing their structures, rather than of analyzing their relationships to the environment.

You cannot observe and record everything; an evaluation of the importance of an observation can only be made within a framework and if that does not yet exist, you are liable to make mistakes. In 1888 when Heinrich Hertz (1857–1894) was attempting to produce the first radio waves, he did not think that the size of his lab or the color of the paint on its walls were relevant to his experiments; he knew from James Clerk Maxwell's (1831–1879) theory of electromagnetism that radio waves were likely to exist, but he could not know that—while the color of the paint was not significant—the size of the lab was because of echoes from the walls.

Theory-laden observation

We can summarize the above discussion by stating that, contrary to a naive concept of induction, in science, observation is *theory-laden*, that is, the theoretical framework within which the observer works is critical to the observations themselves. This is not meant to imply that an observer may not see something that appears in front of his eyes; rather, it means that what the observer notices, remembers, and considers relevant depends on the existence of a theory that guides the observation. Galileo was the first to "observe" the moons of Jupiter and the mountains on the Moon through his telescope. If other people looking through the telescope did not make these observations, it was not because Galileo had better eyesight or because they were being unreasonably obstinate. When you look up into the sky, there are no legends (like the balloons containing conversation in cartoons) that tell you what you see: "Hi, I'm Europa, a moon of Jupiter!" To identify these points of light as stars or planets, you have to know what you are looking for. If this sounds somewhat circular, it is, and it can help explain why scientific advance is difficult.

In Galileo's time, people observed the sky within the framework of the Ptolemaic system (named after the Egyptian astronomer Ptolemy— Claudius Ptolemaeus [ca. 100–170]) that identified the Sun, the Moon, five planets and many fixed stars. This framework had been built up from observations patiently accumulated over centuries. People believed that the heavenly bodies were composed of an incorruptible material that was different from terrestrial material. It was in the *nature* of heavenly bodies to move in circular orbits, while it was in the *nature* of objects on Earth to seek to reach its center. Put yourself in the position of someone handed this new gadget, the telescope, and told to look to the sky. This fellow Galileo points out Jupiter, which you recognize because you are also an astronomer, and asks you to notice a few points of light in its vicinity. For all you know, the points of light could be caused by dust on the surfaces of the lenses or imperfections in their manufacture, or by reflections from other stars or from the candle on your neighbor's balcony. Even in the unlikely case that you have had experience with telescopes, such experience would have been entirely limited to terrestrial objects, and you would have no reason to believe that it performs similarly on heavenly objects. Furthermore, everyone knew that all moving heavenly bodies had already been identified. So, even discounting for jealously and narrow-mindedness, it was entirely reasonable for Galileo's colleagues to be skeptical of his announcement.

Galileo, however, was making these observations from within a different framework. About a hundred years before his time, Nicolaus Copernicus had proposed that the movement of the Sun, Moon, and planets upon the field of the fixed stars could be better explained by assuming that the Earth and all the planets revolved around the Sun, rather than by assuming that the Sun and planets revolved around the Earth. Furthermore, Kepler had discovered that a description of planetary orbits in terms of ellipses was more successful and elegant than a description in terms of circles. Galileo, a first-class mathematician, did not assume that heavenly bodies were necessarily constructed from a more perfect material moving in perfect circular orbits. When he saw a few extra points of light in the vicinity of Jupiter, Galileo was sufficiently intrigued that he continued his observations for several days, recording the positions of the points of light. Only after proposing and rejecting several alternatives, did he convince himself that the best explanation of his observations was that the points were in fact small objects—moons—orbiting around Jupiter. This explanation could only have been made by someone thinking within the new Copernican framework that dethroned the planets from their previous nature.

Clearly, science must start with observation, but once some initial observations have been made, a circular process takes place. Observations lead to theories, which guide further observations, which influence the theories. The presentation of the process of science as initially and primarily inductive is so oversimplified as to be useless. There are serendipitous discoveries in science, in which observations truly instigate the development of theories, but they inevitably occur to those who have the necessary framework within which to understand the importance of what they are observing.

In 1896, Henri Becquerel (1852–1908) discovered radioactivity in a truly serendipitous manner. He was investigating phosphorescence (materials that glow when exposed to light) and he once left a piece of uranium mineral on top of a covered photographic plate. When the plate was later developed it showed signs of having being exposed; this Becquerel interpreted as resulting from radiation emanating from the mineral. Becquerel certainly did not awake that morning and say to himself: "This morning I intend to observe a new scientific phenomenon." But at that time, Becquerel was thoroughly familiar with the recently discovered x-rays, so he was thinking within a framework that allowed for invisible rays. Therefore, it is not particularly strange that he considered the possibility that uranium ore emitted some mysterious rays. The mere observation of the fogged plate did *not* constitute the discovery of radioactivity. Another person, carrying out observations from a different theoretical perspective, would have thrown the plate out as an experimental error, or blamed his assistant for carelessly developing the plate, or tried to return it to the manufacturer as faulty. Instead, Becquerel began a systematic study of the new rays and was awarded the Nobel Prize for Physics in 1903 for the discovery of radioactivity. (The award was shared with Marie [1867–1934] and Pierre Curie [1859–1906].) Yes, Becquerel was lucky, but he made his own luck by pursuing scientific research within a framework that enabled him to appreciate the significance of a serendipitous observation.

The reliability of observation

The second problem with naive induction as the basis for science is that observation is not reliable; what you see is not always what you get. The entertainment industry is full of very talented, reputable professionals who make their living from the creation of unreliable observations—we call them stunt performers, special-effects artists, and magicians. For some psychological reason, we seem to enjoy having our senses, in particular our sense of sight, fooled, and we are willing to pay good money to the artists who create the action TV shows, science-fiction films, and magic shows that never fail to excite us.

There are people who thrive on creating unreliable observations, from criminals, confidence men and women, who use deceptive manipulations to defraud their victims, to professionals like spiritualists who pray on gullible individuals in distress. Magicians, in particular, are incensed at such people whom they believe sully the reputation of their honorable profession. Magician James "the Amazing" Randi (1928–) writes of spoonbender and telepathist Uri Geller (1946–): "[I]f he's using divine power, he's doing it the hard way."[1]

Thus far we have seen that it is not correct to portray the work of a scientist as one who straightforwardly collects observations. Observations can be unreliable; for example, the Moon looks larger on the horizon than it does when it is high in the sky, but that is purely an optical illusion. Observations can also be unreliable because of the limits of technology. For example, if planets revolve around the Sun rather than around the Earth as predicted by Copernicus, the apparent diameter of Venus and Mars should change with the relative positions of these planets with respect to the Earth. But these changes were not observed when Copernicus first proposed his theory and this was taken as evidence against the theory. It was not until the use of telescopes for astronomical observation that reliable measurements of the diameters of these planets could be made and the predictions of the theory confirmed. Nevertheless, observations are the basic stuff from which science is made, and scientists make every effort to ensure that observations are reliable. Great scientists are those who are able to look in novel and imaginative ways at observations that are potentially available to everyone.

Observation is something that has to be learned, because fruitful and reliable observation can only be done within a framework. This is why it takes years of training to become a radiologist and why they are required to qualify first as physicians before specializing in radiology, although it seems that they are performing a simple task, just looking at images. A radiologist must interpret a fuzzy two-dimensional image in terms of an underlying three-dimensional reality, so she must be intimately familiar with human anatomy and the various pathologies that can occur.

Theory-laden observation applies not only within science. A book called *What Great Paintings Say* analyzes a selection of famous paintings both in terms of the details of the scenes portrayed as well as in terms of the underlying social and cultural milieu in which they were painted.[2] After reading the book, when I looked at one of those paintings, I "observed" a great deal more that I would have otherwise, although the physical images would be identical in both cases. Far from diminishing the artistic experience by destroying the "holistic" encounter, reading the book served to increase the enjoyment of the experience by drawing attention to aspects that I would not otherwise have observed.

Observation with instruments

Most of us accept that "seeing is believing," or even that "to see is to know." The relationship between perception and knowledge has been studied for centuries and I will not pretend that I can summarize the various positions. Let us assume for the sake of argument that if you can see something then you can obtain true knowledge of that thing (subject to the caveats mentioned in the previous sections). Some people hold the converse of this statement, namely, that if we can't see something, then we cannot obtain true knowledge about it. The famous nineteenth-century physicist Ernst Mach (1838–1916) (who gave his name to the *mach number* for the speed of sound) refused to acknowledge the existence of atoms because he couldn't see them.

Suppose now that you are watching TV and that you "see" an object like a rock on the surface of the Moon or Mars. Do you obtain true knowledge about that object? I am not talking about ridiculous claims of conspiracy theorists that the Apollo moon landings were faked by NASA; instead, I am trying to evaluate what it means to "see." Well, sunlight reflected from a rock on the surface of the Mars does not enter your eyes directly. The reflected light impinges on an electronic element in a camera where it is converted into a digital signal that is stored in a computer and then transmitted through further electronic circuitry to the Earth. On the Earth, the data representing the picture undergoes computerized enhancement, and is then transmitted to your TV set by another long chain of electronic processing, possibly being sent back into space to be retransmitted by a satellite or converted into light to go through a fiber-optic cable and converted again back into an electronic signal at the end of the cable. Only then, does your TV convert the signal into light that you "see." So we are not really *seeing* the rock.

Very little modern scientific work involves direct sensory experience. Almost invariably, observations and measurements are mediated through chemical markers, electronic signals, and computers. In a trivial sense, thus, modern scientific observation is theory-laden, because you have to understand and accept long chains of reasoning that justify relating the observation to an underlying natural phenomenon. Alan Chalmers (1939–) reports his unease when he supervised a physics experiment in a school lab. The students were measuring the relationship between electric current and magnetic fields, but he knew that the ammeter used to measure the current was based on precisely the phenomenon that the students were supposed to be demonstrating.[3] Of course, such an experiment would never be accepted by a practicing scientist who would insist that the current be measured using an instrument not based on the magnetic phenomenon being investigated.

Normally, however, we don't worry about the use of instruments, neither in everyday life nor in scientific research. We know that TV cameras, communications satellites, fiber-optic cables, and TV sets faithfully reproduce the images of our favorite soap operas, so there is no reason to doubt that the theory underlying these systems suddenly doesn't work because it is transmitting an image of a rock on Mars. Similarly, most instruments used in experimental scientific research have been found through extensive use to give reliable observations, and only a novel instrument will be an object of suspicion until its reliability has been confirmed.

In 1981, Gerd Binnig (1947–) and Heinrich Rohrer (1933–) invented an instrument called the *scanning tunneling microscope (STM)*; a few years later, the *atomic force microscope (AFM)* was invented by Binnig, Christoph Gerber (1942–), and Calvin Quate (1923–). These ingenious instruments are capable of producing images of individual atoms and molecules. They are even capable of manipulating individual atoms, as was demonstrated when Donald Eigler (1953–) and Erhard Schweizer (1957–) of IBM Almaden labs integrated a nice publicity stunt into their research by using an STM to spell out the letters *IBM* using 35 atoms of the element xenon:[4]

Ask yourself now: Would Ernst Mach—who did not accept the existence of atoms because they could not be empirically experienced—now agree that atoms exist? Or would he claim that the underlying theory required to support the claim that STMs show images of atoms is so complicated that we still cannot claim to have "seen" atoms and thus verify their existence?

In fact, the evidence for atoms was already so overwhelming that these visualizations of individual atoms did not contribute to the acceptance of the atomic theory. Instead, the interest was in the instruments themselves and in their potential for the scientific investigation of materials and their engineering applications.

From observation to theory

The next step in the inductive-deductive method is to develop a theory. In chapter 2 we will analyze the concept of scientific theory in depth, but for now it is sufficient to define a theory as a set of concepts and laws based on those concepts that enable you to explain and predict phenomena. For example, the theory of gravitation uses the concepts of distance, mass, and force, and a law relating the gravitational force between two objects to their masses and the distance between them.

So what do you do with a bunch of facts obtained from observation? Enter them into a computer and print out a theory? Laws as such are relatively easy to develop, because we humans seem to have an irresistible tendency to see patterns in the natural world and to generalize from a pattern to a law. Most superstitions started this way: someone was startled a couple of times by a black cat, fell down, and injured himself, so he generalized to a law that black cats bring bad luck, and innumerable generations of black cats have suffered ever since. Developing correct scientific laws is a totally different matter. It takes tremendous effort to identify important concepts and to develop laws that are based on these concepts. That is why science took so long to develop and that is why the pioneers of science are justifiably looked upon as geniuses.

Galileo is admired, not for dropping balls off the Leaning Tower of Pisa, but for being the first to correctly define the concept of acceleration and to realize its central place in a theory of motion. Anyone can observe that an object falls faster if it is dropped from a greater height, but not everyone can develop the mathematical formula that relates the speed at which an object falls to the height from which it was dropped. The most straightforward generalization from the observations is to conclude that the farther an object falls, the faster it travels, or in other words that velocity increases *in proportion to distance*. It was Galileo's genius that led him to identify the concept of acceleration, defined as the change in velocity *in proportion to time*, and that furthermore, to propose the law that states that the acceleration of a falling body under the influence of gravity is constant.

For all major scientific advances, we can usually identify one scientist who came up with precisely the correct concepts and laws. Yet each of them was working within a milieu of intense effort to understand the scientific problems of their day. Many scientists tried to solve these problems, and their contributions were often significant, but we ascribe the advances to the scientists who had the insight to formulate the crucial concepts. Furthermore, these scientists performed extensive calculations, observations, or experiments until they were personally convinced of the correctness of their theories. It is the publication of concepts together with the convincing arguments, calculations, and observations that constitute a scientific advance.

A scientist doesn't collect observations or perform calculations because her job description requires it; being a scientist is not like being a census taker trying to earn a bit of extra money by going door-to-door and collecting facts. A scientist collects observations because she believes that the observations are relevant and will help solve a problem, confirm or refute an existing theory, or develop a new theory. A scientist formulates laws that generalize observations and integrates the laws into theories because she passionately believes that nature can be explained through scientific theories. Performing the "right" observations, experiments, and calculations demands creativity, because there are many ways of doing these activities, and they are all interconnected with one another. It is this creativity that makes science interesting rather than boring.

Prediction and retrodiction

Finally, we come to the third and fourth steps of the naive inductive-deductive method: deducing consequences from the theory and checking these predictions by further observation and experimentation. Deduction is complementary to induction: instead of generalizing from a set of observations to a theory, the general theory is specialized to a particular prediction about a particular experiment or observation. Johannes Kepler performed induction when he generalized astronomical observations into laws expressed as simple mathematical formulas for the orbits of planets. Subsequently, these generalized laws (more precisely, the laws of gravitation and mechanics developed by Isaac Newton [1642–1727]) can be used deductively to predict or to retrodict a specific planet's position at any specific point of time. A *prediction* is the specification of what will happen, whereas a *retrodiction* is a specification of what did happen.

Prediction is, of course, highly useful. Given the initial data concerning a rocket—its position, direction, thrust, and so on—we can precisely predict the future position of the rocket. Dozens of astronauts have entrusted their lives to predictions made from Newtonian mechanics,[5] as well as from theories concerning the performance of rocket engines. Closer to home, every time you fly you are implicitly trusting a range of scientific theories that predict mechanical properties like the strength of the aircraft body, aerodynamic properties like lift and drag, the thermodynamic properties that govern the performance of the engines, and the electromagnetic properties used by the radar and communications systems. These predictions are so successful that the overwhelming majority of flights reach their destinations safely, and accidents are attributed to human error in the design or operation of the aircraft, not to a failure of the underlying theories.

The term retrodiction means "predicting" the past. The concept is not as trivial as it may seem. Obviously, any theory must be consistent with observations that have been performed in the past. But we demand more from a theory—we demand not only that it be consistent with a currently existing set of observations, but also that it be consistent with all new observations, even if they concern events that happened in the past. For example, Newtonian mechanics can be used to predict eclipses of the Sun or the Moon, but it can also be used to retrodict eclipses that occurred in the past. If a reliable historical source is uncovered that reports a solar eclipse, it must be retrodictable by the theory.

Retrodiction is essential if theories are to be developed for the historical sciences. Many aspects of evolutionary biology and geology involve time scales of millions of years that we can barely imagine, much less observe or re-create in a laboratory. Nevertheless, these fields are well-established sciences because they are able to give precise retrodictions of phenomena that occurred in the past. Retrodiction is not just tinkering with a theory until it fits the extant observations. Theories for historical sciences are quite specific in their retrodictions and if evidence is found that something occurred that contradicts a theory, the theory will have to be modified or rejected.

Cosmology, the study of the origins of the universe, is a historical specialty in physics, because any theory in cosmology can be practically used only for retrodiction. (Predictions about the future of our solar system and our galaxy a few billion years from now cannot be confirmed or falsified.) Modern cosmology is based on the *big bang* theory proposed by George Gamow (1904–1968), which claims that the entire universe was created by the explosive expansion of a primeval point. The big bang theory is attractive because it accurately retrodicts the relative amounts of chemical elements that are actually found in the universe. The theory also predicted the existence of background microwave radiation that would be identical in all directions. The subsequent discovery of this radiation by Arno Penzias (1933–) and Robert Wilson (1936–) provided strong evidence that the big bang theory is correct.

Deduction

Deduction is the use of logical reasoning to obtain new statements from existing ones. For example, given the statements: (a) If you touch a hot stove, you will receive a burn, and (b) you touched a hot stove, we can deduce that (c) you received a burn. The justification for the conclusion is that the logical principle we used to carry out the deduction is *sound*; this means that if the premises (a) and (b) are true, then the conclusion (c) is true. The study of deductive systems goes back to the Greeks, continued through the Middle Ages, and in modern times has become a thriving field of research in mathematics. It is possible to formally define what is meant by a deductive system, what it means for a statement to be true and to prove that deductive systems are sound, that is, any statement inferred within the deductive system is a logical consequence of its premises: if the premises are true, so is the conclusion. While formalized deductive systems are usually only studied by professional researchers in the field of mathematical logic, mathematicians and scientists routinely use informal versions of such deductive systems in their reasoning. It is comforting to know that such reasoning is sound. (Deductive logic is discussed in more detail in chapter 11.)

Universal theories

The reason for the importance of deduction in science is that scientific theories are *universal*. Newton's theory of gravitation claims that *all* pairs of objects in the universe attract each other with a force given by a single law expressed by a mathetical equation (see the sidebar) that enables you to calculate the force given the masses of the objects and the distance between them. Since the theory is universal—a force exists between *all* pairs of

---

Newton's Theory of Gravitation

Newton's theory of gravitation includes a mathematical law $F = Gm_1m_2/r^2$ enabling you to calculate the force F between two objects whose masses are m_1 and m_2, separated by a distance r. G is a constant, that is, a property of the universe that can be experimentally measured.

---

objects whatever their masses and at *all* distances of separation—we can use deduction to obtain useful results. We do not need to modify the theory for different uses, we only have to "plug-in" the correct values into the law, though as any student of high-school physics knows, it can be quite difficult to figure out what the correct values are.

When you build a shelf to hold your TV set, you can use the theory to deduce the force of gravity between the TV set and Earth, and then design a shelf that is sufficiently strong to produce an equivalent upward force. If there were a separate theory governing the force between *your* TV set and the planet Earth, you would have no need to deduce anything: you could just look up the force given by the particular theory governing your TV set and the Earth. Of course a science based on separate (and presumably different) laws of gravitation governing each pair of objects in the universe would be so unwieldy as to be totally useless.

Similarly, medicine makes sense as a science because theories of biology and chemistry apply universally to all people. Medical science is nowhere near as precise as physics, so that diagnosis and treatment retain a statistical element (see chapter 10) that requires judgment and experience. However, there is enough universality in the anatomy and physiology of humans, so physicians confidently treat new patients based upon what they learned from textbooks and from clinical experience with previous patients. Medical textbooks and clinical guidelines often have detailed charts describing a process of deduction: "given symptom A, perform test B; if the result is C, treat the patient for D, otherwise, perform test E," and so on. Textbooks do not consist of lists of specific people, and for each one, a list of medical conditions and medications that will work for them.

It is an essential characteristic of scientific theories that they be as universal as possible, otherwise, there would be no point in doing science. Albert Einstein (1879–1955) once said that the most amazing thing about the world is that it is understandable. What he meant was that there is no a priori reason why a few universal theories expressed in concise mathematical equations should govern the universe. Universality is what makes science applicable, and thus different from art or music. You can't "deduce" anything from Ludwig van Beethoven's Ninth Symphony; in fact, you don't even have to appreciate it, and you may not if you grew up in a culture with a different musical tradition. But you can deduce from Newtonian mechanics how to build a shelf or a bridge or how to design a spacecraft, and the same laws are used in all cultures for building shelves and bridges and for designing spacecraft.

Confirmation and falsification

Deduction also provides a way to confirm or falsify a scientific theory. It is risky to propose a universal theory, because once proposed, anyone can deduce (specific) consequences from it. By trying to land a spacecraft on the Moon, you are conducting an experiment that tests the theory of Newtonian mechanics. If the test succeeds, you have evidence that confirms the theory. Suppose, however, that the test failed: Although the spacecraft performed flawlessly, it missed the Moon by thousands of kilometers. Such a failure would supply evidence that falsifies the theory.

There is an inherent asymmetry between *confirmation* and *falsification*. You can never truly confirm a theory, because even if you run a million successful tests, you can never be sure that the next one will also succeed. Furthermore, you can always construct an ad hoc theory that fits a finite number of experiments, for example, the six experiments that tested landing a spacecraft on the Moon. But that is not the same as a universal theory that purports to predict and retrodict all motion not just on Earth but everywhere in the universe. For this reason, scientists tend to avoid phrases like "the experiment proves the theory" and prefer to speak in tones that seem to demonstrate a lack of confidence, using wimpy phrases like "I am led to believe that ..." or "the experiment provides evidence that ...." (See chapter 3 for more on the terminology of science.) You may passionately believe in a theory, but there is always the possibility that the next experiment around the corner will falsify it.

Suppose now that an experiment shows that some prediction (or retrodiction) of a theory, even a single one, is wrong. Let us assume that the deduction of the prediction is sound, and that the experiment has been correctly performed (the equipment is working, the measurements are precise, and all external influences have been accounted for). We must then conclude that the theory itself is wrong, because we deduced a false statement. We say that the theory has been falsified. Science is a risky business that is quite "unfair" to the scientist, because one experiment can falsify a well-loved theory while a million successful experiments cannot absolutely confirm it. (The concept of falsification is rather more involved than this; scientists do not abandon a theory just because of the results of one experiment, but in general it is true that a scientific theory is always in danger of falsification because of experimental results. See chapter 4 for a more thorough discussion.)

During the two centuries after the 1687 publication of Newton's opus *The Mathematical Principles of Natural Philosophy* (usually known as the *Principia*, from the first word in the Latin title), physicists used his theories of gravitation and motion to predict and retrodict the movement of all the bodies in the solar system. The most famous predictions were those that led to the discovery of the planet Neptune on the basis of apparent deviations in the orbit of Uranus, and later the similar discovery of the planet Pluto. Each such success provided confirmation of Newtonian mechanics, to the point that many believed that the theory was *absolutely* true.

Only one blemish remained and it was hoped that it could be rubbed out in the future: The orbit of the planet Mercury deviated slightly from the retrodiction. Even after taking into account the effect on the orbit of Mercury by the gravitational attraction of the other planets, a minute deviation remained: 43 seconds of arc during an entire *century*. (This deviation is little more than the apparent diameter of the Moon as seen from Earth.) Nevertheless, this deviation is now regarded as having falsified Newton's theory of gravitation, which has been replaced by Einstein's theory of general relativity that successfully retrodicted this deviation. Centuries of confirmation were wiped out by the measurements of one pesky planet! Today, Newton's theory is regarded as totally superseded by Einstein's, although Newton's simpler theory is still used, since for most practical purposes the theories are computationally equivalent (see chapter 13).

Linear thinking

One of the strangest criticisms of scientists is that they engage in "linear thinking," whereas others engage in "lateral thinking." (Presumably, "lateral" is superior or at least more satisfying than "linear," though why this should be so is not explained.) The charge derives from a total misconception of the process of science as a literal implementation of the naive inductive-deductive method. A scientist wakes up in the morning, observes some phenomena, induces a theory and then, after lunch, deduces the consequences of the theory and confirms them through experimentation. Any deviation from this linear routine is presumed to be unscientific. Well yes, scientists do make observations and they do perform experiments, and yes, scientists do deduce specific consequences from general laws, but the process is not "linear" and certainly not routine.

Take again the example of Galileo's thought experiment that demonstrates that the motion of falling bodies does not depend on their mass. Surely, hundreds, if not thousands, of people had observed projectiles disintegrate in the air, or chamber pots dropped out of high windows with the result that the pots and their, uh, contents hit the ground together. Galileo's result should have been trivially obvious to the most linear thinker. But it wasn't. The universe does not come with a set of help screens, listing the steps you must follow in order to make discoveries or to formulate scientific theories. Theories tell you what observations might be interesting and how the observations might be interpreted, while observations buffet theories from all sides. The most important advances are those that identify concepts and principles, and these come from intense mental effort, moving back and forth between theory and observation. Frequently, thought experiments are decisive in identifying the essential concepts, but these clearly involve "thinking outside the box," as the teenage Albert Einstein did when he wondered what it would be like to ride upon a beam of light.

The accusation of linear thinking can justifiably be leveled at many science textbooks, especially at the high-school level, which are full of exercises that demand nothing more than setting up the correct equations and solving them, or carrying out the step-by-step instructions in a lab manual. The actual practice of science is quite different, because when you obtain a result from an experiment or a calculation, you can't ask the teacher for the right answer or look it up in the back of the book. Theoreticians don't just plug numbers into formulas; instead, they have to construct models of physical phenomena or perform simulations on computers. They must find the correct abstractions and decide upon simplifications that make computations tractable without seriously harming the validity of the theory. Experimenters must not only devise new techniques, but they must be able to prove that confounding factors are not polluting the results. Fortunately, educational activities are being developed that attempt to provide students with a taste of both the frustration and the satisfaction of working out a solution to a scientific problem.

In chapters 4 and 6 we will describe more sophisticated attempts to define *the* scientific method, though naive acceptance of these proposals is just as mistaken as a naive acceptance of the inductive-deductive method. The scientific process is a complex web of activities and it is doubtful that it can be described by a simple method or set of rules. However, before continuing the discussion on the methodology of science, we will turn to the central concept of science, the scientific theory.

The Nature of Scientists: The Biographical Vignettes

Between chapters we will give short biographical vignettes of some of the most important pioneers of science. This section contains a few general remarks by way of introduction.

Scientists are popularly looked upon as "nerds" or "geeks" with few interests outside the lab or office. A glance at the biographies of famous scientists reveals that they have a few things in common, but otherwise, they are a diverse bunch of characters who could populate a TV soap opera if given the chance.

Great scientists were not necessarily good students in school, but what they do seem to have in common is an intense curiosity about the world and the ability to learn by themselves. If you see a high school student spending long hours passively listening to music or watching TV, you can be certain that you have not met a future winner of the Nobel Prize. But an accomplished musician or an avid reader in many subjects might become one, even if today she is not interested in her high school chemistry class. With the advent of monolithic digital watches, we have probably lost one of the main diagnostic tools for prospective scientists: an interest in disassembling and reassembling clocks!

In the final chapter, we discuss some prognostications concerning the future of science; in particular, we warn that the past will not necessarily repeat itself. This also applies to drawing lessons from the biographies of scientists. The day is long gone when a talented amateur like Michael Faraday (1791–1867) could make important discoveries using homemade apparatuses. To make a new discovery in experimental science requires sophisticated and expensive equipment such as spectrographs, centrifuges, vacuum pumps, lasers, space telescopes, radioactive materials, and deep-sea drilling platforms, and the construction of an experiment and the analysis of the data require extensive knowledge of the underlying theory. In theoretical science, you may need to thoroughly understand graduate-level mathematics just to be able read about current research.

While the biographical vignettes include the illustrious trio of Isaac Newton, Charles Darwin, and Albert Einstein, biographies of modern scientists have been included. Do not draw too many historical lessons from the success of an untrained scientist like Darwin. Without doubt, future scientists will have to study hard, obtain graduate degrees, and join universities and research institutes.

2

Just a Theory:
What Scientists Do

What a theory is not

First let us reject a definition of the term *theory* that probably comes from watching too many detective dramas on TV:

> A theory is a plausible story that is consistent with a set of facts.

Here is an example, starting with a set of facts:

> A fire, which started at approximately 21:00, caused extensive damage to an office. It was determined that many of the cables interconnecting components of the computer had been disconnected before the fire, and that there were traces of a flammable chemical residue.

Now for the "theory":

> The computer powered itself up from standby mode after a power failure. A cat, who was sleeping on the computer, panicked and jumped off into the tangle of cables behind the computer. Some cables ripped loose, upsetting a bottle of cleaning fluid that had been used to clean the monitor. The cleaning fluid ignited from sparking in the loose cables, starting the fire.

Clearly, the theory is a plausible story and it fits the facts. By the end of the TV program, you will surely be told whether this is the correct theory or not. If not, you can easily dismiss it as "just a theory."

Next we must dispose of *teleology*, the attribution of purpose to natural phenomena. Humans are almost certainly the only creatures who ask "why"-questions. We want to know why we are here on Earth, and we want to know the purpose of any action that we do or that is done to us. Aristotle (384–322 BCE), the foremost philosopher and scientist of the ancient world, believed that nature had a teleological aspect, which he called the *final cause*. Aristotle's theory of motion claimed that it is in the nature of terrestrial objects to seek their natural place, namely the center of the Earth. A stone thrown into the air falls back because it attempts to return to its natural place.

Modern science explicitly and emphatically rejects teleology. Physics can describe the trajectory of a falling stone in great detail, but it never attempts to attribute desire or purpose to stones. Biology can describe the evolutionary processes that brought our species *Homo sapiens* onto the face of the Earth, but it has nothing to say about why we are here, nor even if our existence has any purpose whatsoever. Nevertheless, evocative teleological terminology is often used, deepening the confusion of what science is all about. For example, a biologist might say that a species *has adapted* to an environmental niche, implying that the species decided to adapt or strove to adapt. Of course, science claims nothing of the sort. Adaptation is simply the outcome of a process of reproduction amid competition and does not require a decision or intention on the part of any member of the species.

Definition of a scientific theory

Here is a definition that captures the essence of the concept of a scientific theory:

> A *scientific theory* is a concise and coherent set of concepts, claims, and laws (frequently expressed mathematically) that can be used to precisely and accurately explain and predict natural phenomena.
>
> A theory should include a mechanism that explains how its concepts, claims, and laws arise from lower-level theories.

Let us go through the terms that appear in the definition one by one, using Isaac Newton's theory of gravitation as an example.

Concise and coherent

Newton's theory of gravitation makes the claim that there exists an attractive force between every two objects in the universe. It uses the concepts of force, mass, and distance, and the relationship among them is expressed in a single mathematical law given by the equation on page 17. This theory is concise, consisting of one claim, three concepts, and one mathematical law. Even if we add to this count Newton's three laws of mechanics that are needed to apply the theory of gravitation, this is still an impressively concise theory.

Newton's theory is also coherent. Let me explain this by way of a comedy skit I once saw on TV. One comedian starts with an expression of his personal philosophy of ethics: "Goodwill on Earth; peace to all men." His counterpart then starts listing people from professions who—in his opinion—are so unsavory, that they should be exempted from this coherent statement: tax auditors, soccer referees, and so on. (Most of us would probably add spammers to the list!) The first comedian had proposed a concise and coherent principle of ethics, but the second comedian did something that would not be looked upon favorably in science: he destroyed the coherence of the principle by tacking on a large number of ad hoc exceptions. A "theory" of ethics should not include a special case for tax auditors. A scientist feels uncomfortable if special cases are treated by ad hoc modifications of an otherwise coherent theory; instead, special cases should be shown to arise as a prediction of the theory itself, perhaps by taking account of some external factor that was ignored in the original presentation of the theory.

Newton's theory is coherent: it describes a force between objects that can be computed from elementary properties of the bodies themselves—their masses and the distance between them. More importantly, gravitation is universal, applying to *all* objects in the universe. A single concise and coherent theory can explain the motion both of a terrestrial pendulum and a heavenly planet, with no ad hoc rules for special cases. A good rule of thumb for diagnosing an activity as pseudoscientific is the existence of ad hoc explanations: "my telepathic powers aren't working *today* because of a force field emanating from the hostile talk-show host." There are no "bad-gravity days" and there are no days when your TV set stops working because electromagnetic waves feel hostility. Your TV set may experience interference from other electromagnetic waves, but these can be described and measured.

The importance of conciseness and coherence cannot be overemphasized. It is not an exaggeration to state that if a theory cannot be imprinted

---

Science on a Sweatshirt

Maxwell's equations are:

$$\epsilon_0 \oint \mathbf{E} \cdot d\mathbf{S} = q \qquad\qquad \oint \mathbf{B} \cdot d\mathbf{S} = 0$$

$$\oint \mathbf{B} \cdot d\mathbf{l} = \mu_0 \left(\epsilon_0 \frac{d\Phi_E}{dt} + i \right) \qquad\qquad \oint \mathbf{E} \cdot d\mathbf{l} = -\frac{d\Phi_B}{dt}$$

and Schrödinger's equation is:

$$-\frac{\hbar^2}{2} \sum_{j=0}^{N-1} \frac{1}{m_j} \frac{\partial^2}{\partial x_j^2} \Psi(\vec{x}, t) + \mathcal{V}(\vec{x}, t)\Psi(\vec{x}, t) = i\hbar\frac{\partial}{\partial t}\Psi(\vec{x}, t).$$

---

on a sweatshirt, it is not likely to be accepted (sidebar). James Clerk Maxwell's theory of electromagnetism, which explains familiar phenomena such as radio waves and magnetic compasses, consists of four equations. The structure and behavior of atoms and molecules can be derived from one equation of Erwin Schrödinger's (1887–1961) theory of quantum mechanics.

All the strange results of Einstein's special theory of relativity are consequences of two simple principles:

- All laws of physics are the same for uniformly moving observers.

- Light propagates in free space with a speed that is independent of the motion of the source of the light.

Concise and coherent is not the same as "simple and obvious." Decades of research were required in order for the scientific environment to develop to the point where Maxwell, Einstein, and Schrödinger could formulate their theories. Decades more of research were required in order to work out the consequences of the theories, and to perform the experiments and measurements that provided evidence that these theories are correct. It can take years of graduate study for a student to master these theories. But the theories themselves can be expressed concisely and coherently.

Prediction

Science is not alone in its claim to be able to predict the future. Pseudosciences like astrology claim to predict the future, as do religions. But science insists that predictions be precise and accurate. *Accuracy* is a statistical concept that refers to the success of an explanation or prediction. Physics has been able to predict eclipses of the Sun with complete accuracy. Eclipses occur when they are predicted and they do not occur when they are not predicted. It is this accuracy that supports the claim of the underlying theory of Newtonian mechanics to be a successful scientific theory. *Precision* is a measure of the exactness of a measurement. Astronomers are able to predict the exact time at which an eclipse will start and end to a precision of a few seconds, and exactly where on Earth it will be visible to a precision of a few kilometers.

Deep down, scientists' hearts are warmed by precision much more than by accuracy (sidebar). An accurate prediction can occur by chance, or it

---

Precision

The theory of electricity led to an equation containing a constant ϵ_0 called the *permittivity* of free space, which was experimentally determined to be 8.85419×10^{-12} (in units denoted by C^2/Nm^2), and the theory of magnetism led to an equation containing a constant μ_0 called the *permeability* of free space, which was experimentally determined to be 1.25664×10^{-6} (in units denoted Tm/A). ϵ_0 and μ_0 appear in Maxwell's equations (page 26), which can be used to predict that electricity and magnetism interact to create electromagnetic waves such as light, which will travel in free space at a speed c given by the equation $1/\sqrt{\epsilon_0 \mu_0}$. Performing this simple calculation using the experimentally determined values of ϵ_0 and μ_0 predicts that $c = 2.99792 \times 10^8$ m/s, and this is precisely the speed of light obtained in measurements.

---

can be the result of an ad hoc theory or an unidentified factor in an experiment. Suppose that you predict that in ten tosses of a coin, ten heads will come up. Suppose further that you perform such an experiment and ten heads do, in fact, come up. Your accuracy is amazing but less than heart-warming, because, when we do the calculation, we realize that the outcome is not all that unlikely. In 1 out of $2^{10} = 1024$ trials, this result is to be expected. If all the inhabitants of a middle-sized city of one million people performed the experiment, about one thousand of them could be expected to obtain ten heads in a row by chance alone.

A precise prediction, experimentally verified to six significant figures, is something entirely different. You can't help but feel a sense of wonder and awe that science has truly captured how the world works. Physics overflows with precision. Theoretical predictions, worked out from equations dreamed up by Newton in a Lincolnshire apple orchard or by Einstein in a Bern coffee house, have been painstakingly verified to incredible precision, decades or centuries later when experimental techniques became available.

Either science is basically true or we are the victims of an enormous cosmic joke.

The precision of a scientific theory stands in stark contrast to the imprecise predictions of pseudosciences. We are all familiar with the wimpy predictions of fortunate tellers and astrologers: "You will meet an interesting person within the next three days." But you can always define and redefine "interesting" as you see fit: You may interpret the prediction as meaning that you will finally meet your soul mate tomorrow or the day after, but clearly, a tax auditor is also "interesting" by any reasonable definition. To give another example: "According to her horoscope, this baby will grow up to be ambitious, yet amiable." But most people are ambitious and quite a few are amiable as well, so even if the prediction turns out to be accurate, it is hardly precise. On the other hand, if a horoscope could predict a baby's height and weight at age eighteen to within the nearest gram and millimeter, or perhaps predict the exact scores that she will obtain on her college entrance examinations, the prediction would be precise and therefore truly impressive.

Explanation

Humans are notoriously curious. Children go on and on with questions like: "Where does the Sun go to at night?" "Why does sugar disappear when you mix it in your coffee?" "What is thunder?" You might characterize scientists are those humans who have not lost their childhood curiosity, and who continue to search for explanations of natural phenomena.

Explanation is directly related to prediction. If you have an explanation for droughts, it may enable you to predict if another one will occur this year. An explanation for a disease may point to a possible cure or to methods for preventing the disease.

Explanation, a central goal of science, is also a goal of systems of belief. But there is crucial difference between scientific explanation and other forms of explanation. Before the advent of modern science, explanations were *supernatural* and *teleological*. A supernatural explanation is one that goes beyond the natural world for reasons and mechanisms. The variety of supernatural explanations is immense and they can be used to explain the occurrence of any phenomenon. A drought must have been caused by the anger of a god disillusioned with the evil actions of the residents of a region, and a disease must have been caused by the sins of the individual. The problem with supernatural explanations is that they are vacuous. A supernatural entity can be used to explain anything, both a phenomenon and its absence, so it lacks any explanatory or predictive content. Teleological explanations have also been rejected by modern science, which seeks to describe the structure and functioning of the universe, without attributing purpose or volition to natural objects.

Mechanism

Now we come to the most delicate part of the definition of a theory. We have shown that a theory is primarily instrumental: it consists of a concise and coherent set of concepts, claims and laws that enable us to explain and predict phenomena, preferably by calculation, or if not, by some formalization such as the classification systems used in biology and geology. A scientific theory answers "what"-questions, and, by design, avoids answering "why"-questions, taken to mean "for what purpose." There is, however, an intermediate level of questions, "how"-questions, alternatively, "why"-questions, taken to mean "by what means." To take a child's simple question: "Where does the Sun go at night?", Newtonian mechanics can answer by saying that this is an illusion, and that in point of fact, the force of gravity causes the Earth to move around the Sun, so that the Sun illuminates only one-half of the Earth at any one time.[1] Furthermore, we agreed that science would not try to answer the question: "Why does gravity exist?" But surely we should ask the question: "By what means does gravity cause a force to act across such enormous distances?" In other words, a theory should propose a *mechanism*, not just a description of a phenomenon or a rule of calculation.

What would a mechanism for gravity look like? It might explain that the attractive force between two objects is caused by particles exchanged between them, or by waves propagated from one to another. In effect, we are asking for a reduction of the theory of gravitation to an underlying theory (a concept we shall treat in detail in chapter 9). Just to say that gravity "happens" is not scientifically satisfying, anymore than the statement "God wills gravity to exist" is scientifically satisfying.

Here I feel compelled to reveal an embarrassing little secret of physics: there is no fully adequate mechanism that explains gravitation! Both Newton's theory of gravitation and Einstein's theory of general relativity that ultimately replaced it have been extensively tested and found to be excellent theories, in terms of their ability to explain and predict. Newton never suggested any mechanism to explain the gravitational force, and the mechanisms of Einstein's theory have never been empirically verified. Yet Newton's theory is adequate for designing spacecraft and roller coasters, while Einstein's is adequate for designing the GPS system (sidebar). Thus

---

Relativity and the GPS

The *Global Positioning System (GPS)* enables us to obtain an extremely precise fix of our location anywhere on Earth. GPS works by measuring the time it takes for signals from several satellites to reach the instrument, and then using the known position of the satellites to compute the location of the receiver. According to Einstein's special theory of relativity, the atomic clocks on the satellites run *slower* than they do on the Earth (by 7.2 microseconds per day) because the satellites are moving so fast relative to the Earth. According to his general theory of relativity, the clocks run *faster* (by 45.9 microseconds per day) because the force of the Earth's gravity is smaller at the distant satellite than it is for us on the surface. The two effects do not cancel out and a correction factor is used when broadcasting the time signals. Thus, when we use the GPS on airplane flights and wilderness treks, we trust our lives to the success of Einstein's theories, even though the mechanism has not been verified.

---

people trust their lives to theories whose explanatory mechanisms are not fully understood. In fact, theories of gravitation are so well established that if a discrepancy is noted in the orbit of a satellite, it is attributed to local changes in the mass of the Earth (or the Moon or a planet), rather than to any defect in the theories.

Historically, mechanism is the last aspect of a theory to be developed, because it depends on lower-level concepts that are even further from being explained than the theory itself. Modern chemistry developed through decades of research before quantum mechanics was able to provide a mechanism. Maxwell believed that his electromagnetic theory was to be explained as waves in an all-pervasive ether. The mechanism of the ether is long dead and gone, replaced by the photon concept proposed several decades later by Max Planck (1858–1947) and Albert Einstein, but Maxwell's equations are alive and well.

Newton himself was profoundly embarrassed by his inability to provide a mechanism for the force of gravity, as shown in the following famous passage from the *Principia*:

> Hitherto we have explained the phenomena of the heavens and of our sea by the power of gravity, but have not yet assigned the cause of this power. ... But hitherto I have not been able to discover the cause of those properties of gravity from phenomena, and I frame no hypotheses; for whatever is not deduced from the phenomena is to be called an hypothesis; and hypotheses, whether metaphysical or physical, whether of occult qualities or mechanical, have no place in experimental philosophy [i.e., science]. ... And to us it is enough that gravity does really exist, and act according to the laws which we have explained, and abundantly serves to account for all the motions of the celestial bodies, and of our sea.[2]

Perhaps Newton hoped that a mechanism would eventually be found. In the meantime he is saying that we should be satisfied with a concise and coherent theory that accurately predicts and explains motion.

The problem that faced Newton was this: his theory claimed that any two bodies in the universe attracted each other with a force given by a simple formula (page 17). There is no limitation on the distance between the bodies and the presumption is that the force is transmitted instantaneously. "Action at a distance" seems to have more in common with magic and superstition than it does with science. René Descartes (1596–1650) and his followers vehemently objected to Newtonian mechanics and proposed alternate theories based on a medium that transmitted gravitational force between two bodies. The dispute died down because the Cartesians could not supply evidence for the existence of the medium, and the supporters and followers of Newton were extraordinarily successful in explaining and predicting natural phenomena using the theory of gravitation.

The fact that Newton's theory of gravitation is more than adequate as a means to predict and explain demonstrates that we don't reject a theory out-of-hand just because there is no mechanism. Einstein's general theory of relativity is somewhat better in terms of providing a mechanism for gravitation: gravitation results from the deformation of space, and the theory predicts that gravitational fields exist which transmit the force at the speed of light. The fields give rise to gravitational waves and particles that are called *gravitons*, analogous to the electromagnetic waves and photons that come from electromagnetic fields. However, since the force of gravity is extremely weak, experimental evidence for gravitational waves and gravitons is hard to obtain; only certain astronomical objects like quasars and binary stars emit gravitational waves strong enough to be measurable. Currently, the evidence is controversial, so we must live with the embarrassment of risking our lives on a theory whose mechanism is not fully understood.

Evolution is a theory

Let us now evaluate evolution to see if it is "just a theory." First we need a statement of *the theory of evolution by natural selection.* Here are the basic principles of the theory:

- Inheritable variation exists in organisms.

- Organisms reproduce more than the environment can support. If a variation positively or negatively affects reproductive success in the environment, the proportion of the population having this variation will increase or decrease, respectively.

- Variation can become so great that new species can arise. (Individuals that can breed with each other are said to belong to the same species; when individuals in a population can no longer interbreed they are considered to belong to different species.)

That's all that the theory says.

Anyone can see that variation exists (visit a dog show if you are in doubt) and that variation is inheritable (ask a dog breeder if you are in doubt). Simple calculations show that unchecked breeding will lead to a population explosion. If you start with one pair of dogs and double the population every generation, in 33 generations—less than a century—there would be more dogs than people in the world, and in about 260 generations—about 500 years—there would be more dogs than elementary particles in the universe. The theory of evolution by natural selection claims that since populations of a species of organism will necessarily be limited by hunger, disease, or predation, any variation that affects reproductive success will necessarily change the relative proportion of the organisms with that variation, leading eventually to speciation.

Clearly, the theory of evolution by natural selection is concise, consisting of just a handful of principles. Furthermore, it is a coherent theory expressed using a related set of concepts (inheritance, variability and reproductive success), and it applies consistently to *all* living organisms (plants, animals, bacteria, and fungi).

A theory must not only be concise and coherent, it must also explain and predict natural phenomena, accurately and precisely. Let us start with the explanatory power of evolution. Charles Darwin did not set down evolution as an arbitrary theory that sprang from his imagination and then set out to check if it could explain anything. Historically, theories usually emerge because a scientist is looking for a concise, coherent, precise, and accurate explanation for a perplexing set of observations. The reader is referred to Darwin's book, *On the Origin of Species by Means of Natural Selection*, in order to appreciate the massive amount of observations that he accumulated and used to justify his theory.[3]

Darwin, like many others, was perplexed by the diversity of life and the adaptation of organisms to their environment. He tried to understand the geographical distribution of species, the relationships between different species in neighboring areas, and the vast number of fossil species, especially the unknown ones he discovered in South America.

The most important problem that needed to be explained was the adaptation of each organism to its environment. This means that the structure and function of each organism are just what is needed for successful survival. A trivial example is the giraffe's long neck for feeding on vegetation growing high off the ground. Giraffes are adapted for the environment in eastern and southern Africa where they are found, and they are not adapted for the arctic tundra of northern Canada, where they are of course not found. The prevailing ideas concerning evolution at the time held that proto-giraffes stretched their necks to feed on such vegetation and that the stretched necks were then passed on to their descendants. Darwin's genius was in articulating a principle called *natural selection* to explain adaptation. Evolution has nothing to do with the actions or volitions of specific individuals; rather, it is caused by what happens to entire populations. Those animals within a population who had longer necks would be better suited to survive and breed in times of famine, and eventually their offspring became speciated as giraffes.

The theory of evolution by natural selection has proved extremely successful in explaining aspects of the structure of living organisms, including extinct plants and animals. Indeed, the theory is considered to be the most important unifying principle in the science of biology. Evolution explains anomalous findings, such as why whales breath air, rather than extract oxygen from water: Their ancestors were land-dwelling mammals who readapted to living in water, though continuing to breath air. It is possible to construct a precise phylogeny—a description of the evolutionary relationships among all species—though not all biologists agree on the details because different pieces of evidence may support different conclusions. For example, did birds and mammals have a common ancestor that in turn had a common ancestor with crocodiles, or did mammals branch off and only then did birds and crocodiles speciate from a common ancestor. Evolution is a dynamic field of study and scientific arguments may be fierce, but the overall picture—that evolution by natural selection is responsible for the diversity of life and the adaptation of organisms to their environments—is not in doubt.

What about prediction? Can evolution accurately and precisely predict phenomena that are experimentally verifiable?[4] Sometimes this is actually possible. The variation and adaptation of viruses and bacteria is so rapid that evolution can be observed in the laboratory. Evolution predicts that if an agent kills off most, but not all, of a pathogen (disease-causing organism), then those that remain are precisely those variants of the organism that will live longer and thus will be more likely to reproduce in the hostile environment. This rapid evolution explains why it is extremely difficult, if not impossible, to create a vaccine against ailments as innocent as the common cold and as menacing as AIDS. By the time you create a vaccine against one variety, the virus has already evolved into another. Other serious public-health problems have been created by the evolution of the organisms responsible for diseases such as malaria and tuberculosis, which were once easily treatable but are now difficult to treat because mutant pathogens have developed resistance to existing medication.[5]

In general, however, evolution is a historical science, in which retrodiction is much more important than prediction. There is now an extremely large body of phenomena that evolution can explain and we retrodict that any new discovery will be consistent with current explanations (sidebar).

Living Fossils

A particularly satisfying example of a retrodiction of the theory of evolution is to be found in the concept of *living fossils*. In the final chapter of the *Origin*, Darwin wrote:

> Species and groups of species which are called aberrant, and which may fancifully be called living fossils, will aid us in forming a picture of the ancient forms of life.[6]

A living fossil is an existing life form that has remained relatively, but not totally, unchanged throughout the ages, because it is well adapted to some evolutionary niche. The most famous example is the *coelacanth* was first found in South Africa in 1938. The coelacanth is a fish that was known only through fossils dating from 360 through 80 million years ago, yet the structure of the "new" fish was extremely similar to the fossilized remains. This discovery were not of an "aberrant" monstrosity arbitrarily created, but of organisms that fit directly in the known phylogeny, and that had existed for millions of years, although they are only distantly related to most modern species of fish. The discovery of coelacanths provides convincing evidence that fossils really are the remains of ancient living animals and plants, and not just an aberration put upon the Earth to test our faith.[7]

Contrary to popular opinion (including Darwin's!), the best way to falsify evolution is not to claim that some feature of a living thing cannot have evolved because it is too complex. Even if we do not have sufficient evidence to describe the evolution of the feature in detail, we can always wait for further evidence to turn up, either from a fossil or from a better understanding of a current living thing. The statement "I don't see how X could have evolved" simply means that *you* cannot see how X evolved, not that X could not have evolved. Instead, to falsify evolution, you must discover an organism that is totally inconsistent with the entire body of evolutionary biology.

For example, suppose you found an animal with the shell of a turtle, the teeth of a crocodile, the feathers and flight of a bird, and which gives birth to live babies and suckles them with milk. Evolution predicts that such a creature could not exist within the phylogeny that describes life as it exists and existed on Earth. (Of course, it could have existed if the course of evolution had been different.) Evolutionary theorists might find an explanation, or they might modify the theory so that it could explain the existence of this animal, but in the meantime, you could plausibly claim to have falsified evolution.

Mathematics and mechanism in evolution

The theory of evolution by natural selection as propounded by Darwin was originally purely descriptive. It was concise and coherent, offering explanations and predictions that in time proved to be both accurate and precise. Unlike Newton's theory of gravitation, however, evolution was not described in mathematical laws. Like the theory of gravitation, it also lacked a mechanism. Darwin showed that mutations and adaptations occurred, but he was not able to give mathematical laws nor was he able to describe a mechanism for variation and heredity. Darwin himself was almost certainly aware of these difficulties in his theory, because he had been influenced by the quantitative aspect of Thomas Malthus's (1766–1834) treatment of population increases.[8]

The key to the mathematization of evolution came in 1909 with the rediscovery of Georg Mendel's (1822–1884) laws of heredity. During the first half of the twentieth century, research into genetics formed the basis of a mathematical formalization of biology called *population genetics*. It became possible to use assumptions or empirical results concerning the distributions of genes, the effect of genes on an organism, the probability of mutations and so on, in order to develop mathematical laws that would accurately and precisely predict changes in the populations of species. (See

---

The Hardy-Weinberg Equilibrium

Suppose that there are two alleles (variants) A and a of a gene. Most organisms have two copies of each gene, one inherited from each parent, so there will be three genotypes: AA, Aa, and aa. Suppose that they occur initially in the percentages of P, Q, and R, respectively. Then the gene frequencies can be computed as $p = P + (Q/2)$ for A and $q = R + (Q/2)$ for a, because Aa contributes equally to the frequency of both the A and the a gene. Under the assumptions that mating is random, that there is no selection, and that the population is essentially infinite, the population will stabilize at a ratio of p^2, $2pq$, and q^2 for AA, Aa, and aa, respectively. For example, if a population is composed of 25% of AA, 25% of Aa, and 50% of aa, then:

$$p = 0.25 + (0.25/2) = 0.375, \quad q = 0.5 + (0.25/2) = 0.625.$$

The population stabilizes when $0.375^2 = 14\%$ of the group are AA, $2 \cdot 0.375 \cdot 0.625 = 47\%$ are Aa, and $0.625^2 = 39\%$ are aa.

---

the sidebar for a simple law of population genetics.) This synthesis of evolution with genetics, called the *modern synthesis*, was responsible for increasing the stature of the theory of evolution.

By the middle of the twentieth century, the theory of evolution had matured, but there was no mechanism to explain what brought about the phenomena so well explained by the theory. The explosive rise of *molecular biology* has provided this mechanism in amazing detail and depth. There is a mechanism to explain inheritance: Chromosomes separate to form gametes (reproductive cells) and then combine to form zygotes (fertilized eggs). There is a mechanism to explain variability: Errors in the transcription of DNA randomly arise or are caused by environmental agents such as radioactivity. There is a mechanism to explain how variation in the genotype (the genes) causes variation in the phenotype (the expression of the genes in an organism): An organism is built from proteins that are synthesized under the control of RNA copied from the DNA, so differences in DNA translate into differences in the organism itself. The mechanism of evolution is thus found in the theories of biochemistry and molecular genetics.

Evolution is a living science. There are still unknowns, such as how the biochemical basis of life came into being; there are still heated disagreements among scientists on important aspects of evolution, such as whether evolution works gradually or in spurts. Nevertheless, the theory of evolution more than fulfills all of the requirements of scientific "theoryhood," even more than the theory of gravitation. To brand evolution as "just a theory" is the finest compliment one can confer on it!

Are *creationism* and *intelligent design* theories?

The theory of evolution by natural selection has always had to contend with objections from people who believe that the theory contradicts religious doctrine, in particular, the doctrine of special creation, which is the belief that God created each species individually. The Bill of Rights appended to the US Constitution prohibits the establishment of religion by the government, and this is interpreted to preclude the teaching of religion in public schools. In an attempt to introduce the teaching of special creation into public schools, the doctrine is now expressed in a form that purports to be a scientific theory. This doctrine, called *creationism*, and more recently *intelligent design (ID)*, holds that the existence of complex living organisms on Earth can only be explained by the existence of an intelligent being who designed them. The inference of a creator from the existence of complexity is call the *argument from design*, and has been used for centuries to justify religious beliefs. The most famous exposition of the argument from design was given by William Paley (1743–1805) in 1802. He inferred the existence of a creator of the universe by analogy with the inference from the complexity of a watch to a designer of the watch.

Let us now analyze ID within the framework of our definition of a scientific theory.[9] Intelligent design is certainly concise, consisting of just the assertion that living organisms have been designed. It hardly makes sense to talk about ID as coherent, because there are no interrelated concepts, and furthermore, it is totally ad hoc. If you ask why a giraffe has a long neck, the answer is that it was designed that way. ID is essentially a total failure of the imagination; just because *you* do not see how something could have evolved, doesn't mean that it didn't.

Let us consider an updated version of Paley's example of design: my favorite toy, a handheld personal digital assistant (PDA). If you found one lying on the street, you would certainly marvel at its complexity and its exquisite adaptation, and infer the existence of a designer. While it is strictly true that some engineer in Silicon Valley designed the particular model and some worker in China assembled the specific object that I own, I can use my knowledge of computer technology to show how its design evolved.

If I were to use the concepts of ID, I would infer the existence of a designer by marvelling at the perfect adaptation of the LCD screen, the touch screen, the handwriting recognition software, the graphical user interface, and the operating system with its ability to keep time and to communicate with other computers. Clearly, if any one of these "organs" were missing, the PDA would be totally useless; for example, handwriting recognition would be totally useless without a touch screen. But just as clearly, I *know* that each of these organs evolved for entirely different reasons. The development of LCD screens was driven by weight considerations for portable computers; touch screens were used in industrial and military environments where an ordinary keyboard would not survive; handwriting recognition is useful for reading checks and collecting data from documents; graphical user interfaces were developed in order to make computer software more user-friendly; and operating system development was driven by efficiency demands in the use of mainframe computers. The magnificent adaptation of the PDA cannot be used to toss away decades of evolution of computer technology. Similarly, the fact that we cannot perfectly reconstruct the history of biological evolution does not mean that the existence of a magnificent organism *requires* that it was specially created.

Needless to say, there is no formal or mathematical content to ID—a "theory" that explains everything, explains nothing, and predicts nothing. ID cannot explain why millions of species were created and then became extinct. Even more importantly, it cannot explain "mistakes" in the design of living organisms such as us. The eye is the classical example of an organ that is supposed to be so complex that it must have been designed, yet our eyes have a serious design flaw in that the blood vessels and nerves are placed on the surface of the retina instead of behind it. Ophthalmologists spend much of their time trying to treat the degradation of sight that can result from this flaw. If you bought a video camera with such a defect, you wouldn't hesitate an instant before returning it to the store. Now it is not as if the designer didn't know any better: Cephalopods (squids and octopuses) have eyes that are "correctly designed."[10]

Here is another example. There is a long discussion in chapter 13 of Darwin's *Origin* in which he asks why there are no *Batrachians* (frogs and toads) native to oceanic islands, though these animals thrived when they were introduced by humans. His answer was based on experiments showing that both adult frogs and their embryonic forms (eggs and tadpoles) are killed by contact with salt water. Therefore, since frogs evolved in bodies of fresh water on some continent, the mechanisms that he proposed for the migration of species to islands (floating on driftwood or attached to the feet of oceanic birds) were not effective in this case.

What can ID contribute to this scientific question? Exactly nothing. All it can state is that it was the will of the designer not to place frogs in the Azores. That statement may or may not be correct, but it certainly has no scientific content whatsoever. You cannot deduce useful predictions from ID, because there are no limitations on the designer. So if we actually did find that animal with the shell of a turtle, the teeth of a crocodile, etc., (see page 36) it would not invoke any wonder, because the designer simply decided to create such a creature.

Finally, in the absence of a detailed description of how the designer works, there is no mechanism.

Is evolution a religion?

Deep down inside, creationists are aware that they are promoting religion, not science, because they ascribe the same to evolution:

> Thus, evolution is surely a religion, in every sense of the word. It is a world view, a philosophy of life and meaning, an attempt to explain the origin and the development of everything, from elements of galaxies to people, without the necessity of an omnipotent, personal, transcendent Creator.[11]

But evolution only claims to be a theory that explains how the forms of living organisms are modified as time goes on; it stands on its internal conciseness and coherence, and on its external support by the evidence. It is not a religion, and most biologists would be surprised to find out that they are using a theory that is really a worldview that explains "the origin and development of everything."

Creationists extrapolate the term evolution from a specific theory within biology to a blanket term covering the origin of everything:

Over the ages, if evolution is true, primeval particles have evolved into molecules and galaxies, inorganic chemicals have developed into living cells, and protozoans have evolved into humans, so there must be some grand principle of increasing organization and complexity functioning in nature.[12]

The big bang theory for the origin of "molecules and galaxies" could be true or false, independently of whether the theory that "protozoans have evolved into humans" is true or false. In fact, the evidence in favor of the theory of evolution is far more massive and convincing than the evidence for the big bang theory.

Finally, the nonscientific status of creationism is clear from the terminology used:

In any case, if the term "creationism" is used, then "evolutionism" should be used correspondingly.[13]

Scientists generally do not call themselves after the theories they are studying, but after the topic of study. Thus we have physicists and chemists, but not "gravitationists" and "electromagneticists" and "quantum mechanicists."[14]

* * *

A scientific theory is the starting point for research: We are interested in explaining more and more phenomena, exploring the theory's ramifications to develop predictions that can be confirmed or falsified, and delving deeper into the underlying mechanisms. A scientific theory is dynamic, subject to controversy and modification as more empirical evidence is obtained and as other theories are able to provide a mechanism for the theory. The theory of evolution by natural selection is not "just a theory"; it *is* a theory, and a successful one, according to the scientific meaning of the term. Intelligent Design is simply a dead end; it does not *deserve* to be called a theory.

ISAAC NEWTON: FROM ASCETIC GENIUS TO ACERBIC MANAGER

Isaac Newton (1642–1727) was probably as close to being a classical nerd as any famous scientist. For over thirty years, Newton spent all of his time performing research in his combination office, laboratory, and residence at Trinity College of Cambridge University. He never married and may have been celibate his whole life. His position at Cambridge and an inheritance from his mother supplied his monetary needs. Newton was profoundly religious and wrote more on the history of Christianity and on Christian theology than he did on any other subject.

Newton's father died before he was born—a small, weak baby who was not expected to live. His mother remarried shortly afterward, but young Isaac never got along with his stepfather. The family was able to afford an education, and Newton boarded with a pharmacist near the school, perhaps to escape his home environment, as the school was not far from his mother's house. Even though Newton was not a promising schoolboy, we can detect the beginnings of his interest in science. He would build models and tools, and he spent hours with his landlord learning to prepare medicines, which pharmacists had to mix by hand in the days before there were manufactured pills. Newton retained a deep interest in chemistry and alchemy, though his scientific achievements were in physics and mathematics.

Newton attended Trinity College in Cambridge from 1661 to 1668, and at age 27 was appointed Lucasian Professor of Mathematics in 1669. His tutor guided his studies of the classics, but he had to teach himself mathematics and physics. Newton's most significant work was performed during 1666, known as his "Miracle Year." Cambridge University was temporarily closed because of a plague and Newton returned to live at home. His insights into motion and gravitation that constitute Newtonian mechanics date from this period, as do many of the detailed calculations that provide the evidence in favor of these theories. The story about the apple is quite probably true. Newton may have been watching the Moon near the horizon, just as an apple fell in his line of sight. He understood that the Moon was in fact falling just as the apple did, but that its horizontal motion was so great that it never reached the Earth.

Although the young Newton had solved the foremost problems of the physics of his day, and established the methods and tools of mathematical physics, he did not publish his opus, the *Principia*, for another twenty years. This was partially to allow him to work out detailed consequences of the theory and partly to avoid premature exposure of his results to his rivals. Newton engaged in bitter feuds with Robert Hooke (1635–1703) and Gottfried Wilhelm Leibnitz (1646–1716) over scientific precedence. His oft-repeated remark: "If I have seen farther, it is by standing on the shoulders of Giants," may not have been what it is assumed to be, a sign of modesty and an acknowledgement of his intellectual debt to Galileo and others, but a cruel, sarcastic barb aimed at the "vertically-challenged" Hooke.[15]

In addition to his pioneering work in physics, Newton spent much time on activities like alchemy that in retrospect were a waste of time. He never published his massive writings on theology, because he came to believe in Arianism, a version of Christianity that holds that Jesus is not divine. If these heretical views had become public, he would have been stripped of his positions and social standing.

At age 54, another side of Newton appears: Newton the reclusive nerd became Newton the administrator and government official. He began work at the Royal Mint in London, receiving the top job of Master of the Mint within a few years. He chased counterfeiters mercilessly, to the gallows if need be, and recruited an extensive intelligence network among London's criminals. Can you even imagine a movie script that calls for the foremost physicist of our day to hang out in smoky pubs and on foggy street corners waiting to meet his informants?!

Newton was highly respected during his lifetime: He was knighted and was elected several times to Parliament, though he remained inactive as a member. In 1703, Newton was elected president of the foremost British scientific institute, the Royal Society. He reorganized the society, placing it on a sound financial and administrative footing, and spent several years searching for a suitable building to house the society and supervising its renovation.

Until the end of his life at age eighty-four, Newton continued to revise his scientific works.

3   Words Scientists Don't Use: At Least Not the Way You Do

Scientists deal in *facts*, don't they? Scientists *prove* their theories; that's why they *believe* in science, isn't it? Actually, scientists tend to avoid these terms, or, at the very least, they use them informally with meanings that are quite different from their day-to-day meanings. In the previous chapter we have examined at length the difference between the colloquial use of the word *theory* and its scientific meaning. To help you better understand the nature of science, we will continue to examine the meaning of words as used by scientists.

Fact

The term *fact* is normally reserved for an observation or explanation that is absolutely true. But no scientist would claim that anything is absolutely true. At most, they would claim that the *preponderance of evidence* points to the truth of the observation or explanation. Even a simple observation statement like "The ball I threw hit the ground" is not absolutely true, because one can never be sure that the observation is not the result of an illusion. How many times have you seen a pretty young woman sliced in two on a stage without calling the police? If "seeing is believing," you should have called the police to report the brutal crime committed by the magician. But as we discussed in chapter 1, observations are theory laden: You need an appropriate theoretical background in order to observe what your eyes see. When you see a magic show, you are observing within the theoretical framework of entertainment by illusion, in which what you see is almost certainly not what it appears to be. A person who knew nothing of magic would almost certainly be horrified by the illusion.

Similarly, when the TV detective asks for "just the facts, ma'am," he is not looking for a comprehensive list of everything the witness saw: (1) "The second window from the left is not broken"; (2) "There is a bowl of apples on the coffee table"; (3) "Light is coming from the chandelier." Neither is he looking for deductive conclusions: "That rat of a business partner of his must have shot him." One the one hand, the detective is interested in obtaining observations that are filtered as little as possible by conjecture (since it is his job to make the conjectures), but on the other hand he is still asking for observations from the theoretical background of the usual inhabitant of the house. So (1-3) above are not interesting, while (4) "I saw my husband lying in a pool of blood in the kitchen with a knife in his back" is an observation that is of profound interest.

People get "stabbed" all the time on stage and live to tell about it, but if your husband is lying motionless in a pool of blood on the floor of the kitchen, the *reliability* of the observation is quite high, and no useful purpose would be served by doubting the observation and considering the possibility that the knife is a fake, that the red stuff is ketchup, and that your husband is play-acting, even though such an explanation is possible.

Scientists are in a deeper quandary, because they are making observations and proposing explanations about unknown phenomena. You are doing science when you are trying to observe or explain a phenomenon that is not fully understood; if you already knew the answer, there would be no point in performing a calculation or carrying out an experiment. Clearly, when dealing with the unknown you are likely to be misled by illusions. When Galileo first observed the moons of Jupiter, all he saw were a few spots that could easily have been caused by dust on the lenses or imperfections in his eye. It was only by keeping careful records of these spots over days and weeks that he was able to interpret them as bodies rotating around Jupiter. He also observed the twinkling of stars, but we now know that this is an illusion caused by the Earth's atmosphere, not by an observable property of stars.

We can define a fact as an observation backed up by such a preponderance of evidence that no useful purpose would be served by doubting it. Under this definition of fact, "The ball I threw hit the ground" is a fact. There is a roughly analogous definition in the legal system, which requires that the evidence show that the defendant has committed the crime beyond any *reasonable* doubt. If the requirement were beyond *any doubt whatsoever*, the prisons would empty out, because every criminal could claim that he had been abducted by aliens and temporarily replaced by an alien double, or that the witness had seen a magical illusion. There is no way to refute these possibilities absolutely, but no useful purpose is served by doubting the straightforward explanation, namely, that the defendant in the courtroom is the same person the witness saw rob the bank, not an alien and not a computer-generated image. The requirements on the establishment of a fact in the legal system are much less stringent than they are in science. Retrials are rarely granted to convicted criminals, whereas any scientific fact—no matter how venerable and no matter how eminent the scientist who established it—is always subject to new efforts to falsify it.

A scientific observation becomes a fact when there is no longer any reason to doubt it. Thus, the report of the observation and its acceptance by the scientific community are part of what makes the observation a fact. The acceptance comes from an analysis of the report of the experimental or observational conditions; confirmation by independent observers, often using different techniques, is frequently carried out. If there *is* reasonable doubt, that is, if there is a real reason to doubt an observation, there will always be some thesis-topic-hungry graduate student or tenure-questing faculty member who will try to show that the observation was mistaken. Eventually, as time passes, and as attempts to falsify observations fail, while attempts to observe the phenomenon in different situations confirm the original observation, it becomes to be accepted as a fact. There is always the possibility that all these scientists have misinterpreted their measurements or deluded themselves, but this possibility becomes less and less tenable as more checking and cross-checking goes on.

Creationists raise the possibility that the universe was really created only about six thousand years ago, and that the geological and fossil record, which shows that the Earth is about five billion years old and that life first appeared over three billion years ago, is an illusion created by God in order to test our faith. There is no way of absolutely disproving that the observations are illusions, but the preponderance of evidence for an Earth existing for billions of years is so overwhelming that no useful purpose is served by doubt. This evidence has been obtained using different methods of radiological dating, paleontological and stratigraphical studies, investigations into changes of the climate, and the Earth's magnetic field. It is a *fact* that the Earth and life upon it have existed for billions of years.

Theories as facts

Not only do observations and measurements become accepted as facts, even theories can also be considered to be facts when no useful purpose would be served by doubting them. Newton's theory of gravitation was a fact, and in a sense still is, because for most practical purposes the preponderance of evidence supports the theory. When astronauts were sent to the Moon, they may have worried about bugs in the navigation software, or short circuits in the computer hardware, or leaks in the propulsion system, but they did not give a second thought to the possibility that the theory of gravitation would turn out to be incorrect on the far side of the Moon. The situation is similar with the theory of evolution by natural selection. The preponderance of evidence supports the theory and no useful purpose is served by doubting it.

It must be emphasized that what we call the preponderance of evidence supports a general theory or a class of observations. The specific details of a theory or the status of individual observations may engender acrimonious disputes among scientists, without causing any of them to doubt the factual basis of the underlying theory. For example, the preponderance of evidence points to the arrival of humans into the American continents over a land bridge between Siberia and Alaska. But scientists wage a vociferous debate as to when this occurred, and estimates range from just over ten thousand years ago to about thirty thousand years ago. The theory of the arrival of humans in America from Siberia can be considered a fact, but the date of that arrival is not. It is possible that new evidence may eventually accumulate to the point that scientists can resolve the debate and come to a consensus, but it is also possible that such uncontroversial evidence may never be forthcoming, so that some particular date for the arrival of humans in America may or may not become a fact.

Law

A scientific *law* is *descriptive*, not *prescriptive* or *normative*. When used in the context of a legal system, a law prescribes a certain behavior that must or must not be done. There is a law that you must bring your car to a full stop if you encounter a red traffic light. This is a totally arbitrary convention, not only in the sense that the law could have specified that you stop when the light is purple, but also in the sense that the entire concept of traffic signals to regulate the flow of cars and pedestrians is just a social construction. Furthermore, although the convention of stopping on red is probably universal, there are local differences in the precise details of the law; for example, some countries permit you to make a right turn at a red light, while others prohibit it. Airplanes that travel in three dimensions have their movements regulated by entirely different conventions, and there is no a priori reason why we couldn't have ground traffic controllers instead of traffic lights. Laws are also used to specify social norms of behavior: Schoolchildren are forced to obey innumerable laws like "don't run in the hallways" and "raise your hand and wait to be recognized if you want to talk." Finally, laws can be "broken," and all societies have judicial institutions for dealing with lawbreakers.

In science, a law is a concise description of a regularity, usually obtained by generalizing from a set of observations.[1] The law of gravitation is simply a formula that describes how the magnitude and direction of forces on two bodies is related to their masses and to the distances between them. The law does not *require* bodies to behave in a certain way, nor does it *tell* the bodies how they *should* behave, and they certainly cannot "disobey" the law.

It is important to distinguish between the scientific concept of law as a generalization, and the social concept of law which is prescriptive and normative. A desire for tolerance in respecting the laws of different social systems must not lead us into the mistake of attributing volition to the entities of science or relativism to scientific laws. Laws of nature are simply description of what *is*; science fiction writers might speculate on what the world would be like if some laws were different, but we can't change the laws.

Proof

Colloquially, you *prove* something by supplying sufficient evidence for a claim to convince someone of its truth. A prosecutor will try to prove the guilt of a defendant. The term *proof* is also used colloquially as in: "He *proved* that he is serious by taking me to meet his parents." When scientists use the word proof—which is rarely—they are talking about one of two things: a purely mathematical or logical demonstration that a certain theory implies a certain phenomenon, or the mass of confirming evidence that justifies accepting a theory. Proof is not a good word to use in science, because it conveys the certainty of the mathematical concept of proof, which is far removed from the scientific concept of evidence supporting a theory.

A mathematical proof is a logical demonstration that a conclusion must follow from a set of premises by rules of inference. Unfortunately, a proof can never establish the absolute truth of a claim, only the truth of a claim relative to a set of premises. Consider the statement: "Newton proved Johannes Kepler's first law that the orbits of planets are ellipses." This statement is perfectly acceptable when interpreted in the following way: Newton gave a mathematical derivation (proof) that *if* two bodies attract each other with a force proportional to the product of their masses and inversely proportional to the square of the distance between them, and *if* the first body [the Sun] is much more massive than the second body [a planet], and *if* the second body is traveling at orbital speed in rotation around the first, *then* the orbit of the smaller body is an ellipse with the larger body at one of its foci. As a mathematical claim this would hold whether or not the Sun and planets ever existed. As a scientific claim, however, it doesn't prove very much of anything, because you have to first establish the truth of all the premises. But you can't establish the truth of the theory of gravitation independently of the observational and experimental evidence that confirms it, and there is no a priori reason why the theory of gravitation shouldn't have a different form.

Newton's proof is interesting because it demonstrates the interesting *mathematical* property that an inverse-square law of attraction entails elliptic orbits. But it is not a scientific proof that the claim is true; rather, it provides evidence supporting Newton's theory of gravitation. Once Kepler had established by deep analysis of many years' worth of observations that the planets do orbit the Sun in elliptical orbits, these observations served as evidence for Newton's theory because they were retrodictions of the theory. As we discuss in chapter 11, it would be fallacious to conclude that this argument—called the affirmation of the consequent—*proves* the truth of the theory of gravitation, but it does rule out many other possible theories and provides significant evidence in its favor.

Belief

Renowned astronomer and science writer Carl Sagan (1934–1996) was often asked if he *believed* in extraterrestrial life. His reply was that it is not a matter of belief but of evidence, and he had yet to see any convincing evidence of its existence.[2] If a scientist uses the word the belief, it is again used within the limited connotation of the preponderance of evidence supporting an observation or a theory. Alternatively, it can be used with a meaning that is almost diametrically opposed to its everyday meaning—a scientist may say that he believes that a certain claim is true (perhaps in the context of a proposal for a research grant), but the claim is "just a belief" with no little or no solid evidence to back it up. The word belief used thus is roughly synonymous with the words hypothesis and conjecture, which are to be preferred. The difference between the two is this: *conjecture* is used when there is evidence for a claim, but the evidence is not sufficient to justify the claim, while *hypothesis* is used for a claim that is assumed for the purpose of investigating its consequences, without any real evidence for its truth. Very often, a scientific claim progresses from hypothesis to conjecture to fact.

Though science is not a religion in the sense of having a canon of beliefs that must be adhered to, nevertheless, there are certain things that most, if not all, scientists do believe in:

- The universe exists.

- The universe functions according to uniform, natural, (usually) mathematical laws.

- These laws can be discovered by observation and experimentation, and they can be incorporated into theories that can explain and predict.

These are not a priori beliefs handed down by a prophet, but beliefs that have emerged from the success of the scientific enterprise. No scientist is examined to ensure that she actually believes these tenets, nor is she excommunicated for not holding them, but it is hard to see why you would devote your life to science if you did not believe them.

The difference between these beliefs and the beliefs of religions is that scientific beliefs are *methodological*, not *propositional*. A propositional belief is one in which you hold that a certain specific proposition is true: "The speed of light is approximately 3×10^8 meters/second ($300, 000, 000$ meters per second)." A methodological belief is one that has no propositional content, but just guides your behavior: "If articles in several reputable, peer-reviewed scientific journals describe experiments showing that the speed of light is approximately 3×10^8 meters/second, then I accept this as a fact and plan my own experiments using this fact." These beliefs are partly a matter of commonsense and partly the result of the success of science, for if the universe did not exist and did not function according to uniform, natural laws that can be studied, there would be no science and hence no scientific belief.

Almost everyone believes these principles in one form or another. You do so implicitly every time you take medication for a headache. If the universe doesn't exist, perhaps your headache doesn't exist either. If the universe doesn't function according to uniform, natural laws, then the active ingredients in the medication may have been efficacious yesterday but not today, since the laws of biochemistry might have changed. If theories can't explain and predict, and if experiments can't offer evidence, then the results of clinical experiments leave you no reason to believe that the medication is likely to have the desired effect.

It is the absence of propositional beliefs that distinguish science from a belief system. There is no proposition of the content of science that is accepted upon belief alone. If, for example, someone would show that the speed of light is about 4×10^8 meters/second instead of the currently accepted measurement which yields about 3×10^8 meters/second, scientists would be puzzled, shocked, and dismayed, but the edifice of science would not crumble. Most religions, on the other hand, require that you believe in specific propositions like the occurrence of Noah's flood. If you do not believe in the religion's corpus of propositions, you might be forced to leave that religious community (or at least you will experience dissonance between your inner beliefs and you outward profession of them).

In chapter 6 we will discuss opposition from postmodernists to the methodological beliefs about science. They look upon science as a subjective, purely human creation that does not necessarily reflect a real universe. For scientists, the belief that the universe is reasonable and understandable is a belief that is reasonable and understandable.

Progress

Progress is a favorite word of politicians: "Together, we shall progress to a better and brighter future." Well, what did you expect him to say? "Together, we shall regress to the evil times of the past." Even conservatives who dream of the good old days will use language reminiscent of progress:

"Let us restore the grandeur of yesteryear for the sake of our children and our children's children." Advertisers are also great believers in progress, offering us a new and improved formula for something that was perfectly satisfactory to begin with. Even when a product maintains its popularity for decades, manufacturers will give the packaging a face-lift in order to progress with fashion.

It is important to distinguish between the use of the word progress to describe the discovery of new knowledge and the invention of new techniques, and its normative connotation as it appears in the rhetoric of politicians. There is no question that science is progressive in the sense that more scientific knowledge exists today than it did a few years or decades ago. We know that a relatively simple theory of gravitation explains the motion of the planets—a piece of knowledge that did not exist five hundred years ago; we know that bacteria and viruses cause many diseases—a piece of knowledge that did not exist as late as one hundred and fifty years ago. You simply cannot "uninvent" something, though under conditions of social turmoil, for example, in the "Dark Ages," knowledge may come to be lost in some parts of the world.

There are many people who are not happy with scientific progress and who are nostalgic for previous eras (but see page 167 before you indulge in too much nostalgia). This does not change the fact that science has progressed in the sense of the accumulation of more knowledge. It is a matter for personal opinion whether this constitutes progress in the sense of granting people better lives. The scientific knowledge about nuclear physics developed in the twentieth century constitutes progress, though many people would not regard this as progress in terms of making them happier. In an implicit recognition of the irreversibility of scientific progress, some people object to scientific research, believing that more harm than good comes from the progress of science.

Anthropomorphism

Anthropomorphism is the attribution of human characteristics to a nonhuman entity. Such language is sometimes used by scientists, but the use is quite innocent and there is no intention that the attributions be taken literally. If a scientist says: "The electron wants to fall, but he finds it difficult to do so because of the electric potential," she is not attributing any human characteristics to the electron. An electron is not a he (or a she for that matter), it doesn't want anything and it doesn't find anything difficult. I am reminded of the riddle: "What is the most amazing technological invention of all time?" Answer: "The thermos bottle: it keeps hot drinks hot and cold drinks cold, but how does it know which is which?!" Any use of anthropomorphic language should be discounted immediately so that you don't fall into the teleological trap that we discussed in chapter 1. Science takes upon itself only to describe phenomena, not to attribute purpose or volition.

Sometimes, writers will get so carried away with mellifluous prose that the result is totally meaningless, but likely to cause damage to the intellectual health of the unwary. Here are some quotes from an article in the intellectual press, where I have emphasized the vacuous anthropomorphisms:

> [T]he universe is not an *indifferent engineering scaffold* but is *steeped in resplendent mystery.*
>
> But rather than the *heartless domain* assumed by contemporary thought, a multiverse would be an almost *preternaturally sanguine place.*
>
> [S]cience is trending away from dispirited views of a *merciless cosmos* toward a new vision of creation as *poignantly favorable to life.*[3]

Even scientists are not immune from temptation!

> String theory has the potential to show that all of the wondrous happenings in the universe—from the *frantic dance* of subatomic quarks to the *stately waltz* of orbiting binary stars, from the *primordial fireball* of the big bang to the *majestic swirl* of heavenly galaxies—are reflections of one grand physical principle, one master equation.[4]

Numbers

Mathematicians give evocative names to mundane mathematical entities, and these names should not be construed as having an intrinsic meaning. These names are important historically, because our understanding of mathematics (beyond arithmetic and geometry, which were well known to the Greeks and others) developed slowly over the past few centuries. In fact, a satisfactory theory of numbers was not given until the late nineteenth century.

There is something to be said for the term *the natural numbers* (0, 1, 2, ...), because they are quite intuitive (although the classification of 0 as a number was not easily achieved). Today we accept *negative numbers* as equally intuitive, though the concept was initially problematical because numbers were associated with counting objects and properties like length and weight. With the advent of analytic geometry and the interpretation of negative numbers as displacements along an axis rather than lengths, they lost their mystique.

The terms *rational* numbers (for example, 1/2)and *irrational* numbers (for example, $\sqrt{2}$) originate in the philosophical system of the Pythagoreans. They attributed metaphysical properties to ratios of lengths so that numbers like $\sqrt{2}$ that could not be so expressed were outside their system of thought. Rational numbers are simply pairs of numbers that obey certain axioms (sidebar). Irrational numbers are not in need of psychiatric help; they are simply *numbers* which are not rational.

Rational numbers

When you learned about rational numbers in high school, a special notation called *fractions* was introduced and rules were given for computing with fractions:

$$\frac{a}{b} \times \frac{c}{d} = \frac{a \times c}{b \times d} \qquad \frac{a}{b} + \frac{c}{d} = \frac{(a \times d) + (b \times c)}{b \times d}.$$

But without the fractional notation, rational numbers are just pairs of numbers (a, b) that are defined to obey axioms like $(a, b) \times (c, d) = (a \times c, b \times d)$. All the concepts of mathematics can be built up by defining sets of natural numbers or arbitrary symbols obeying a set of axioms.

Transcendental numbers (for example, π) did not originate in an Indian ashram; they have an equally straightforward definition as numbers that are not the solution of algebraic equations like the ones you studied in high school.

The square root of -1 ($\sqrt{-1}$), posed similar problems in the conceptual development of mathematics. It was named an *imaginary* number, while ordinary numbers were called *real* numbers, and a pair consisting of one of each was called a *complex* number. The formal definition of real numbers is rather difficult, but they have an intuitive interpretation as displacements along an axis. Similarly, imaginary numbers are not imaginary and the theory of complex numbers is no more complex than the theory of real numbers. Complex numbers are as intuitive for electronics engineers as -100 is for the average person with an overdrawn bank account.

The terminology of evolution

The theory of evolution by natural selection abounds in misleading terminology! Who has not heard the terms natural selection, adaptation, and survival of the fittest? Here we will analyze the terminology of evolution in order to clarify their technical meaning and to free them from the colloquial connotations that they have.

Natural selection: The word *select* tends to be used in a transitive sense, so we assume that there must be a "selector" as well as a "selectee." Darwin chose the word because he drew a deep analogy from *artificial selection* as practiced by animal breeders. A breeder will select a cow who produces more milk over one who produces less, eventually leading to a population of cows more suitable for dairy farming. The endless variety of dogs of all sizes, shapes, colors, and dispositions is clear evidence for the power of artificial selection. The analog in the theory of evolution is that adaptation takes places because variations are "selected for" within environmental niches. The criterion for selection is no longer the set of regulations governing dog shows, but the ability to reproduce in an environment. But the analogy is not complete because there is no breeder performing the selection. A tree top about to furnish nourishment to a proto-giraffe in a drought can hardly be said to be actively "selecting." Selection is something that just happens; no one is doing the selection.

Adaptation: Similarly, there is confusion resulting from the teleological connotation of *adaptation*. An individual proto-giraffe did not (actively) adapt by stretching its neck; instead, in a population of proto-giraffes, animals with longer necks were able to live longer, to reproduce more often and to tend their long-necked calves more effectively. Eventually, the population became (passively) adapted and speciated as giraffes.

Teleology is also related to a view of evolution as *progress* from microorganisms to humans at the pinnacle of life, a view that is attractive because it is consistent with religious beliefs that view the world as created for humans to serve God. It is true that complex organisms evolve as adaptations to highly specialized conditions, but there is no purpose in evolution.[5] Many complex organisms have become extinct and many organisms *regress*, for example, viruses which were originally independent organisms became more "primitive" and totally dependent on their hosts.

Survival of the fittest: There is no question, however, that first prize for misleading terminology must go to the phrase *survival of the fittest*, which, by the way, was not used by Darwin himself in the *Origin*, but by sociologist Herbert Spencer (1820–1903). While technically accurate as a synonym for reproductive advantage, it conveys the totally incorrect connotation that "fit" somehow means physically or mentally or morally fit, as in: "He is the *fittest* man I've ever seen; he must work out everyday in the gym." Or: "He doesn't have the strength of personality that would make him *fit* to become president of the company."

Survival of the fittest has also been construed to involve not just competition but violence as well. In 1850, nine years before the publication of the *Origin*, Alfred Lord Tennyson (1809–1892) published the poem "In Memoriam," which the Victorians took to their hearts because it seemed to promise that people would become better in the future. It contained a line "Nature, red in tooth and claw" that came to symbolize Darwin's theory of evolution. But evolution applies equally to cuddly, bamboo-chomping pandas as it does to the lions who sink their reddened teeth and claws into peace-loving gazelles. Reproductive advantage comes in many forms. In the world of the gazelle, survival of the fittest does not mean survival of those individuals who can slug it out with a lion, but survival of those who sense danger quicker, run away faster, or are more camouflaged within the environment. For every predator for whom survival of the fittest means selection for aggression, there is a prey for whom survival of the fittest means selection for timidity.

The theory of evolution provided a fertile field in which ideologues of all stripes grazed, primarily those advocating "progress." The idea was that if nature can "progress" by evolving something as complex and well adapted as humans, then surely humans can "progress" by evolving their social, political, and economic characteristics. Of course, Darwin's theory based on natural selection is not an appropriate theoretical framework to discuss such issues, because it is neither progressive nor teleological. They were based upon a theory proposed by Jean-Baptiste Lamarck (1744–1829) who claimed that the effort by an organism to improve itself could affect heredity.[6] Larmarck's claim was a reasonable conjecture given the state of knowledge in the early nineteenth century, but it has be totally rejected in favor of Darwin's theory of evolution by natural selection.

Perhaps the darkest travesty of science has been the misuse of the phrase *survival of the fittest* in the form of *social Darwinism* and *eugenics* to justify cruel political ideologies. According to extreme social Darwinists, there is no point in helping the poor or handicapped, because—for the good of the human race—they should be allowed to lose out in the struggle for existence and die off. According to eugenics, "unfit" humans should be prevented from reproducing. Social Darwinism was used to justify the sterilization and murder of those deemed by the Nazis to be politically, ethnically, morally, or mentally unfit.

Social Darwinism and eugenics are unscientific in two aspects. First, they committed the *naturalistic fallacy*, which is the assumption that what is, must be. The fact that the natural environment selects for reproductive advantage does not mean that we, as humans, should be forcibly selecting people according to some preconceived notions. The second nonscientific aspect of these movements was their narrow interpretation of the meaning of fitness. In evolution, this simply means fitness to survive and reproduce, not fitness according to some externally imposed criteria.

Darwin himself never engaged in these speculations, nor did he support these perversions of his theory.

Origins: It is important to emphasize that Darwin did not set out to "explain the origin ... of everything" as creationist Henry Morris (1918–) would have (see the full quote on page 40). All he set out to do was to explain the origin of *species*, that is, why there are so many different life forms, both living and extinct, that are well adapted to their environments. Initially, Darwin did not doubt that life was divinely created (though he became less religious as time went on), but even to a religious person it seemed incredible that the creator would take the trouble to create so many species in obscure places upon the Earth, and even more incredible that he would create many millions of species just to drive them to extinction before humans, the epitome of creation, appeared.

The origin of *life* is an entirely separate issue. Many scientists think that life evolved spontaneously from chemicals in the Earth's environment five billion years ago, and there have been experiments trying to recreate the biochemical basis of life. Theistic evolutionists believe that life itself was divinely created and then species evolved as described by the theory of evolution. Darwin himself preferred not to take sides in this issue, devoting himself to the understanding of the details of speciation by natural selection. The final sentence of the *Origin* quoted in the next paragraph shows how he temporized on this issue, even allowing for the possibility that a "few" life forms were created before evolution took over.

Evolve: Finally—it is more than a curiosity—let us note that the word "evolve" does not occur in the *Origin* except as its final word! The final sentence is, however, sufficiently poetic to be worth quoting:

> There is grandeur in this view of life, with its several powers, having been originally breathed into a few forms or into one; and that, whilst this planet has gone cycling on according to the fixed law of gravity, from so simple a beginning endless forms most beautiful and most wonderful have been, and are being, evolved.

The theory of evolution by natural selection is at its core a set of principles and laws that can explain and predict the development of life forms over time. It forms the basis of modern biology and an understanding of evolution is crucial to an understanding of the life sciences. The word evolution has connotations of teleology that are irrelevant to the theory. Like Darwin, we can look upon the grandeur of this view of life, or we can look upon it with abhorrence. But evolution itself simply happened; the theory itself says nothing about the goal or purpose of evolution, or the lack thereof.

CHARLES DARWIN: HYPOCHONDRIAC REVOLUTIONARY

Charles Darwin (1809–1882) was fortunate enough to be born into a well-to-do family and did not have to support himself. He was a poor student at school (though a talented athlete), and for many years he couldn't "find himself" as we would say today. He studied medicine in Edinburgh but dropped out, and then began to study theology at Cambridge University, preparing for the quiet life of a minister in the Anglican Church. During the years of his studies, Darwin developed an interest in nature, becoming an obsessive collector of beetles. Through attending lectures and meetings at scientific societies, he began to learn the scientific aspects of nature studies. This led in 1831 to an offer of the position of companion to Captain Fitzroy of the ship HMS *Beagle*, which was about to embark on a two-year voyage to South America.[7] Initially, his father objected to this as a waste of time, but was eventually convinced to give the twenty-two-year-old Darwin his blessing and some financial support. This rest, as they say, is history.

The HMS *Beagle* left Plymouth, England, on December 27, 1831, and sailed for the eastern coast of South America, where surveys were conducted from Brazil down to Tierra del Fuego until mid-1834. Then it sailed into the Pacific Ocean to survey the western coast. In the fall of 1835, the HMS *Beagle* visited the Galapagos Islands. In October, it sailed westward, finally landing at Falmouth a year later on October 2, 1836. At every stop along the route of this five-year voyage, Darwin went on expeditions, often for days or weeks at a time. He observed geological formations, and collected and preserved biological specimens, both living and fossil, which he sent back to England. Everywhere he observed that the Earth itself and all its life forms were extremely diverse, well beyond the imagination of a stay-at-home naturalist.

Upon his return from the long voyage, Darwin arranged for his geological and biological collections to be examined by experts, and began to write about his voyage and his observations. By 1842, he had written, but not published, an outline of the theory of evolution by natural selection. Aware of its revolutionary nature, Darwin refrained from publication until he had amassed and analyzed more evidence. Publication of the *Origin* in 1859 was instigated by a paper by Alfred Wallace (1823–1913) proposing a similar theory.

Darwin continued to work on the theory of evolution for the rest of his life. He was primarily interested in the adaptation of species to their environments and carried out many experiments with plants and small animals (barnacles and frogs).

In 1839, Darwin married his cousin Emma Wedgwood (1808–1896) (yes, from the rich family of blue-and-white porcelain fame), and in 1842 bought a country house at Downe, Kent, where he remained for forty years until the end of his life. They had a long, happy marriage with ten children, three unfortunately dying young.

Darwin was unable to work for extended periods of time because of illness. There are many fascinating theories concerning this illness. One possibility is that it was a genuine tropical illness that he contracted during his travels in South America. Some believe that he exaggerated minor problems as a way of avoiding visitors and traveling. Most likely, the illness was of a psychosomatic nature, which leaves a wide field for speculation as to the underlying psychological problem. One popular proposal is that Darwin suffered from guilt feelings caused by his own drift into agnosticism bordering on atheism as the result of his research on evolution. A related conjecture is that he felt truly saddened by the distress this caused his wife who had remained devoutly religious. Finally, some religious people have suggested that Darwin was tortured by God as punishment for promoting atheism.[8]

4   Falsificationism: If It Might Be Wrong, It's Science

In chapter 1, we discussed the asymmetry between confirmation and falsification: a million confirming observations merely provide evidence not absolute truth for the correctness of a theory, whereas a single falsifying observation may destroy a theory. The philosopher of science Karl Popper (1902–1994) based his description of science upon the concept of falsification of theories.[1] First we will discuss the use of the concept of falsification as a means of drawing the boundary between science and nonscience; then we will discuss Popper's attempts to describe a scientific methodology based upon falsification.

Demarcation

The first contribution of falsification to an understanding of the nature of science is in *demarcating* science from nonscience. How are we to know if some activity is to be considered scientific or not? According to Karl Popper, a theory is scientific only if its statements are *falsifiable*, that is, if—at least in principle—some observations or experiments can lead us to reject the theory. Falsifiability is property of a theory that is independent of the truth of the theory; furthermore, a theory can be falsifiable, even if the falsifying experiment is not feasible.

Clearly, Newtonian mechanics is falsifiable, in fact, it is a false theory, because it has been falsified by the many experiments that confirm the predictions of the theories of relativity. However, let us consider the status of the theory before the twentieth century. You can use Newtonian mechanics (together with a basic data like the size and mass of the Earth) to deduce that the escape velocity from the Earth is about 40,000 kilometers per hour. This is the velocity at which a body will no longer fall back to the surface of the Earth, but instead continue onward into space. No one of Newton's generation was in a position to perform an experiment to check this prediction, but the mere fact that such an experiment was conceivable was sufficient to make the theory falsifiable. Today, such an experiment is implicitly performed every time a spacecraft is sent to the Moon or beyond, and these experiments have all been successful. Even missions that failed are attributed to defects in the rockets, not to errors in Newtonian mechanics.

These successful experiments supplied evidence that confirms Newtonian mechanics, but Popper was more interested in the fact that the experiments could have failed. Let us suppose that a failed experiment showed that this prediction did not hold, and that a spacecraft launched at 40,000 kilometers/hour fell back to the Earth's surface, or that an airplane designed to fly at a leisurely 1000 kilometers/hour suddenly darted off into space. Newton, as well as every other scientist, had to live with the possibility that his theory would one day be falsified.

Popper proposed his principle of falsification in reaction to the claims of Sigmund Freud (1856–1939) to have developed a scientific theory of the human mind and the claims of Karl Marx (1818–1883) to have developed a scientific theory of history. Popper claimed that neither psychoanalysis nor Marxism were falsifiable in the same manner as Newtonian mechanics. Exactly how could psychoanalysis be falsified, even in principle? This would require an observation or experiment that would falsify some precise prediction of the theory. Suppose that the theory of psychoanalysis claimed that: "If the patient has a dream of form X, then he is suffering from disease Y." If we found a patient who dreamed X but did not suffer from Y, then we would have falsified psychoanalysis. Popper did not believe that it would ever be possible to formulate and test falsifiable statements in psychoanalysis.

Similarly, Popper did not think that political and economic history could ever become precise enough so that predictions from Marx's theory could be falsifiable. On the surface though, Marxism seems to be falsifiable: Marx claimed that there was a necessary progression from feudalism to capitalism to socialism, and this is a prediction that is falsifiable. In fact, it seems to have been falsified. Russia and China, the two ostensibly Marxist states, passed directly from feudalism to socialism without passing through a capitalist stage, and they are now moving from socialism to capitalism, thus doubly falsifying the theory.

If a "theory" is not falsifiable and thus not scientific, it does not mean that it is uninteresting or not useful. Although psychoanalysis may not be scientific, many psychiatrists believe that it is a useful therapeutic tool.[2] Marxism certainly qualifies as interesting for its huge impact on the history of the twentieth-century.

Degrees of falsifiability

Popper's principle of falsification goes beyond the simple requirement that a scientific theory be falsifiable; he claimed that theories should be judged according to their degree of falsifiability.

> A theory A is *more falsifiable than* a theory B if the predictions of A are more precise than B, therefore offering more opportunities to falsify it.

The classic example is to compare Kepler's Laws with Newtonian mechanics. Johannes Kepler (1571–1630) gave three laws describing the orbits of the planets around the Sun (sidebar). Newton deduced Kepler's laws as consequences from his laws of motion and theory of gravitation. Newtonian mechanics is therefore *more falsifiable* than Kepler's Laws, be-

---

Kepler's Laws

Johannes Kepler worked for Tycho Brahe (1546–1601) who made the most accurate and extensive astronomical observations ever done before the development of modern telescopes. By analyzing the data, Kepler was able to formulate three laws:[3]

- All planets move in elliptical orbits about the Sun.

- If you draw an imaginary line from the Sun to the planet, then the line sweeps out equal areas in equal time intervals.

- For all planets, the ratio of the cube of the distance from the Sun to the square of the orbital period is constant.

It is clear that Kepler's laws are descriptive and apply only to planets orbiting the Sun. Newton's great achievement was to provide a detailed mathematical proof that these laws are consequences of a *universal* theory of gravitation that applies to any object whatsoever.

cause they apply to the motion of *all* objects and not just to a handful of planets. The discovery of a planet with a nonelliptical orbit would falsify both theories, but the discovery of a type of rock that did not fall in accordance with Newtonian mechanics would falsify only Newtonian mechanics and not Kepler's laws, for the trivial reason that Kepler said nothing about falling rocks.

In fact, Newtonian mechanics has been falsified, not just by detecting some unexplainable anomaly in planetary motion that also falsifies Kepler's laws (see page 19), but also by comparing atomic clocks flying in airplanes with identical clocks on the ground, and by measuring properties of elementary particles on Earth. A highly falsifiable theory is to be preferred to one that is less falsifiable, because it is more risky and therefore— as long as it has not been falsified—more likely to be interesting and significant.

The concept of degrees of falsifiability has influenced the social sciences, causing them to enthusiastically adopt quantitative research methods. To say that 40 percent of first-born sons develop a psychological syndrome is certainly more precise than just saying that first-born sons "tend to" develop that syndrome. It is also more falsifiable, in the sense that if an experiment resulted in a different value, say 15 percent, then the theory would be falsified and must be abandoned or modified. On the other hand, both 40 and 15 percent can be interpreted as confirming a prediction expressed as "tend to." The question is to what extent quantitative results are useful in fields like psychology and sociology that deal with the complexity of human behavior. There is no real predictive value in knowing that 8 percent of a population develop some psychological syndrome, unless there is a reliable test for the syndrome and a mechanism that will explain why the result is 8 percent and not 80 percent. Currently, there is a trend to do research in social sciences using the ethnographic methods of anthropology, which are based on observation and detailed analysis of the specific observations rather than upon statistical analysis of populations. These methods make the resultant theories less falsifiable and thus according to Popper less scientific, but the results are often more interesting and useful.

Falsification as a methodology

Karl Popper took the concept of falsification beyond a principle to demarcate science from nonscience. He looked upon the progress of science as a never-ending cycle in which a theory is proposed, then falsified, and then another theory is offered in its place, the new theory itself becoming a candidate for falsification. Falsification thus became the basis of a *methodology* of science. Confirming an existing theory is like treading water; it gets you nowhere. Instead, scientists should attempt to falsify existing theories and then search for theories that solve the problems of these theories.

The following diagram shows how science progresses according to falsificationism:

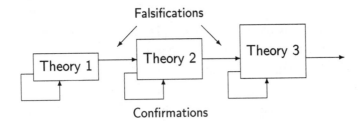

Scientists propose a theory and then perform experiments to check predictions of the theory. If an experiment confirms the predictions, there is no incentive to change the theory. If, however, an experiment refutes the predictions of the theory, the theory is falsified, and scientists must search for a new theory not refuted by the experiment (nor by any previous experiments that had confirmed previous theories). As we describe in the next section, it is rare for a single experiment to be so decisive as to falsify a theory, but eventually the set of observations or experiments that refute the theory can become so overwhelming that scientists have no choice but to consider the theory to be falsified.

As a methodology, falsification has certain advantages over naive induction. A scientist working according to this methodology is guided by a theory she wishes to falsify when deciding what observations to collect and what experiments to perform. Thus falsification, unlike induction, recognizes that observation is theory laden, performed within a theoretical framework. Furthermore, falsification avoids talking about the "truth" of a theory; instead, one theory is simply "better" than another because it has survived more attempts at falsification.

While it is true that scientists do engage in attempts to falsify existing theories, as a methodology falsification simply does not describe what actually happens in the process of science. First, scientists are primarily motivated by the ambition to propose and confirm theories that they themselves believe to be true, not to falsify theories suggested by others. For example, the Michelson-Morley experiment was carried out in 1887 to confirm the existence of ether as a medium for the propagation of light waves, although the experimental results turned out to falsify that theory. Similarly, Arthur Eddington's (1882–1944) experiment in 1919, which showed that light is bent when it grazes the Sun, was carried out in order to confirm the general theory of relativity, and the success of the experiment led to the widespread acceptance of the theory.

A second difficulty with falsification as a methodology is that the validity of a theory is not judged by simply checking experimental results against the theory, but by a complex process of mutual feedback within the community of scientists. The development of a scientific theory requires intense mental effort, so that it comes as no surprise that a scientist may be so emotionally attached to a theory that occasional or even frequent falsifications will not shake him from his belief in the correctness of the theory. Those holding existing theories will not necessarily accept a new theory merely because it explains a few observations that falsify the existing theories.

Is falsification compatible with the history of science?

Let us now examine a historical episode which illustrates that science does not progress according to the straightforward cycle: theory followed by falsification, followed by a new theory that is falsified in its turn.[4]

Albert Einstein published his theory of special relativity in 1905; the theory was intended to reconcile serious discrepancies between Newtonian mechanics and Maxwell's theory of electromagnetism. The laws of Newtonian mechanics allow simple transformations that took account of the relative position and velocity of two objects, whereas Maxwell's theory did not. This was one of the most important dilemmas in physics at the time and several physicists had proposed alternative theories to solve the dilemma. In 1906, Walter Kaufmann (1871–1947) measured the deflection of beta rays in electromagnetic fields. According to the theory of special relativity, the mass of the high-speed electrons that form the beta rays should increase at high velocities. The increased mass should reduce the deflection of the beta rays when an electromagnetic field is applied, and the change in deflection can be calculated.

Kaufmann's result turned out to be closer to the prediction given by an alternative theory proposed by physicist Max Abraham (1875–1922) than to the prediction deduced from special relativity. Furthermore, the prediction was also unfavorable to the theory developed by Hendrik Lorentz (1853–1928) whose equations Einstein had adapted. (Lorentz himself never accepted the theory of relativity, though he remained a good friend and admirer of Einstein.) Lorentz took the news very hard as it seemed to falsify his theory, while Einstein shrugged it off. After the intense effort he devoted to developing special relativity, there was no way that he was going to accept a single result as a falsification. Surely, Kaufmann's measurements must be wrong, as indeed they turned out to be.

According to the principle of falsification, Einstein should have accepted that the theory of special relativity had been falsified. Recall that if we accept the soundness of a deductive system, then if a conclusion is false, the premises must have been false. Therefore, if Kaufmann's result conflicted with a prediction of the theory of special relativity, it is the theory that must be false. So, was Einstein justified in obstinately holding to his theory? And if so, why was he justified in persisting in his claims?

A famous anecdote tells of a student who asked Einstein what his reaction would have been if Eddington's experiment had not supported his theory of general relativity. Einstein replied: "I would have felt sorry for the Good Lord because the theory *is* correct."

The apparent contradiction between simple logic and Einstein's behavior can be explained if we reexamine the notion of deduction from a theory that many of us hold as a result of doing too many high-school physics exercises (sidebar). The answer, of course, is always in the back of the book, or at least in the instructor's manual, and it will be revealed after the exam. Unfortunately, the universe does not come with an instructor's manual and technical support is as hard to get as it is for some software packages. Furthermore, the universe was not created in accordance with modern pedagogical principles, so you cannot run an ideal experiment in which there is an isolated pair of point masses. High-school science experiments often go to great lengths to try to create ideal conditions, but even

---

High-School Physics Exercises

Physics exercises typically give you precisely the amount of data needed to solve them, neither too much nor too little. For example: Determine the mass of the earth from the equation $G = (r^2/m_{earth})g$. (Answer: $m_{earth} = 5.98 \times 10^{24}$ kg.)[5]

---

with our a posteriori knowledge of how the results should come out, they rarely do.

Consider, now, the situation of a scientist like Galileo performing an experiment, or a theoretician like Einstein weighing the effect of an experimental result on his theory. They don't know the result beforehand, they can't check the result in the back of the book and they don't know which factors are likely to obscure the phenomenon they are investigating. A falsifying result negates one or more of the premises, but there is no way of knowing which one. It might be the theory itself or it might be any of a large number of initial conditions that are an essential part of the experiment. These can range from inaccuracies in measurements, supposedly negligible influences that aren't, contamination of materials, and psychological factors like wishful thinking. Experimental science is no less demanding than theoretical science, because the scientist must be able to demonstrate that a result truly confirms or falsifies a consequence of a theory and is not caused by a mistake. Therefore, until Kaufmann's experiment could be checked and rechecked, Einstein was justified in holding on to his theory and proposing instead that the experiment was flawed.

Blondlot's n-rays

A classic example of wishful thinking in science concerns the "discovery" of *n-rays* by French physicist René Blondlot (1849–1930) in 1903.[6] Perhaps envious of the discovery of x-rays by German physicist Wilhelm Röntgen (1845–1923), Blondlot claimed to have discovered a new form of radiation and proceeded to perform research on their properties. However, scientists in other countries could not duplicate his experiments. American physicist Robert Wood (1868–1955) visited Blondlot's laboratory and surreptitiously removed a critical component of the experimental appara-

tus; nevertheless, Blondlot continued to observe his n-rays. Wood's falsification of n-rays was quickly accepted by the scientific community, and
Blondlot never recovered from the blow to his reputation. Here we see an
instance of a successful falsification, but it is actually more complex than
it seems. First, Wood had to remove a critical component of the system,
not simply manipulate some adjustment, so that the falsification could not
be attributed to a minor problem with an initial condition. Second, Wood's
trick would not have had the effect that it did if it had not been done within
a scientific environment of general rejection of n-rays because they had not
been observed in other labs.

There is no question that falsification as description of the methodology
of science is not correct. Certainly, it is not a useful prescription of how
scientists should carry out their research. The question remains: How can
we distinguish laudable stubborn insistence on the correctness of a theory
from pig-headed refusal to abandon an incorrect theory? The short answer
is that there is no a priori way to distinguish between them. The long
answer will be given in chapter 6, where we delve into the processes that
lead scientists as a community to accept one result and reject another.

How did Darwin confrontation potential falsification?

Although Popper's presentation of falsification as a methodology of science does not describe what actually happens in science, falsification as a
concept seems to be central to scientific ways of thinking. All scientists are
aware that confirmation is never absolute and that falsification can come at
any time. If falsifying observations and experiments keep piling up, eventually a theory must be abandoned or modified. A good scientist is one
who attempts to anticipate possible falsifications and preempt them.

Darwin is a prime example. By 1838, Darwin had put together the
basic concepts of the theory of evolution, and by 1844 he had written a
short (230-page!) essay setting out the theory, but his famous book *On the
Origin of Species by Means of Natural Selection* was not published until
1859. Furthermore, Darwin considered the 600-page *Origin* to be a mere
abstract of an extremely long multivolume work. Why did Darwin wait so
long and why is his abstract the size of a modern textbook?

Reading the *Origin* is fascinating, even for those of us who are not professional biologists. The theory of evolution by natural selection is quite

concise and easy enough to explain in a few pages, but Darwin was well aware that it would be difficult to accept:

> Although I am fully convinced of the truth of the views given in this volume under the form of an abstract, I by no means expect to convince experienced naturalists whose minds are stocked with a multitude of facts all viewed, during a long course of years, from a point of view directly opposite to mine. It is so easy to hide our ignorance under such expressions as the 'plan of creation,' 'unity of design,' &c., and to think that we give an explanation when we only restate a fact. Any one whose disposition leads him to attach more weight to unexplained difficulties than to the explanation of a certain number of facts will certainly reject my theory.[7]

Nevertheless, he expected that some readers would be willing to weigh the evidence, and if not, certainly the next generation would do so:

> A few naturalists, endowed with much flexibility of mind, and who have already begun to doubt on the immutability of species, may be influenced by this volume; but I look with confidence to the future, to young and rising naturalists, who will be able to view both sides of the question with impartiality.[8]

Darwin regarded it as essential that he collect and present as much evidence as possible to support his theory.

Darwin spent those decades of "inactivity" performing empirical studies, and carrying on extensive correspondence with naturalists who were studying the plants and animals that he would later use as evidence. He spent eight years of research on *barnacles* (*Cirripedia*), which are crustaceans like shrimp and crabs that attach themselves to rocks, piers, and even ships. Barnacles exist in a vast number of varieties and species, proving to Darwin that natural selection could bring about significant changes in organisms over a long period of time.[9] Here are some photographs of barnacles named after Charles Darwin:[10]

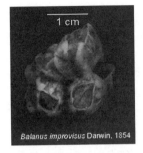

Balanus improvisus Darwin, 1854

Balanus trigonus Darwin, 1854

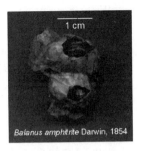

Balanus amphitrite Darwin, 1854

He also carried out an intensive study of the breeding of farm animals and pigeons. Six-hundred pages really do contain too little room to adduce all the evidence that he had collected. The *Origin* is awash with apologies by Darwin for its brevity and with promises that details will be forthcoming in the full treatment of evolution. Darwin revised the *Origin* many times as new evidence became available and to counter specific criticism, but he never wrote the promised complete work.

Darwin claims that he was assiduous in collecting potential falsifications:

> I had, also, during many years followed a golden rule, namely, that whenever a published fact, a new observation or thought came across me, which was opposed to my general results, to make a memorandum of it without fail and at once; for I had found by experience that such facts and thoughts were far more apt to escape from the memory than favourable ones. Owing to this habit, very few objections were raised against my views which I had not at least noticed and attempted to answer.[11]

The second half of the *Origin* contains several chapters in which Darwin explicitly anticipates attempts to falsify his theory of evolution, whether these reservations were actually expressed to him or whether he was pre-empting potential claims against the theory. He counters each of them with an explanation, frequently backed up by experiments that he performed specifically to address a possible falsification. Today we know that many of the difficulties he anticipated can be explained by processes that he knew nothing about, such as molecular biology and continental drift (see chapter 12). Nevertheless, by reading the *Origin*, you can appreciate Darwin's conscientiousness as a scientist who was fully aware of the possibilities that his theory might be falsified.

With the advent of molecular biology, the degree of falsification of the theory of evolution increased. The classification of organisms that was based upon morphology must be consistent with the classification based upon molecular genetics. If we find two organisms—perhaps two primates—who are morphologically very similar but genetically very different, then the theory of evolution by natural selection will be in serious trouble. In fact, not only are we morphologically very similar to the great apes, we are also genetically very similar.[12] The consistency of morphology and genetics has supplied overwhelming support for the theory of evolution, evidence that Darwin could only dream about.

How do creationists confront potential falsification?

Creationists are much less forthcoming on the possibility of falsifying their "theory." Here is an unambiguous prediction written by Henry Morris, a leading creationist:

> Consequently the Creationist predicts that no transitional sequences (except within each created type) will ever be found, either in the present array of organisms or in the fossil record.[13]

Presumably, if such a transitional sequence is found, creationism will be falsified, but their acceptance of this possibility is not explicitly stated for the simple reason that for religious reasons creationists would never be influenced by such falsifications.

An example can show how the falsifiability of creationism is a moving target. Creationists were fond of claiming that evolution was falsified because of the lack of transitional forms between land mammals and marine mammals. They found it impossible to imagine how a mammal could gradually become adapted to marine life and how a partial adaptation could be practical. However, recent paleontological research has uncovered fossils of half-a-dozen species that clearly form a transitional sequence between land mammals and whales. It is now clear that about fifty million years ago *artiodactyls* (hippopotamuses, pigs, cows) and *cetaceans* (whales, dolphins, porpoises) had common ancestors, though the precise relationship is still subject to controversy. *Ambolocetus natans*, a creature whose fossil was discovered in 1992, shows precisely the transitional features that one would expect from a species intermediate between land mammals and whales: it was quite capable of thriving in the boundary between land and water, walking and swimming as needed.[14]

This transitional sequence clearly falsifies Morris's prediction from creationism, but I doubt whether he will abandon creationism and accept the theory of evolution for which the sequence supplies confirmatory evidence. Of course, Einstein also did not abandon relativity at the first difficulty, but this transitional sequence is but one of many, furnishing the preponderance of evidence for the theory of evolution. Until creationists accept that their claims must be falsifiable and show how they could be falsified, creationism cannot be said to be a scientific theory.

* * *

In our definition of a theory, the phrase "precisely and accurately predict natural phenomena" implicitly states that a scientific theory must be falsifiable. If a scientist uses a theory to predict a phenomenon and the prediction turns out to be false, the theory must be false, though as we have seen such a conclusion is not straightforward, because the problem may be with the observations themselves or with the assumptions on the use of the theory. Popper's attempt to construct a methodology from falsification may have been flawed, but the concept is central in the demarcation of science from nonscience. No matter how strong one's convictions, a true scientist will always allow for the possibility that her results may be falsified; if she denies this possibility or refuses to abandon or modify a theory in the face of repeated falsifications, you can be sure that you are dealing with pseudoscience, not science.

ALBERT EINSTEIN: PARADIGM OF A GENIUS

Let us start by refuting a persistent myth. Any time that a student has difficulties in school, teachers and parents often encourage him with the statement that Einstein was not a good student. Now it is true that Albert Einstein (1879–1955) was an unruly and impudent student, and that the regimented and highly formal German school system in the late nineteenth century did not take kindly to his behavior. He even skipped most of his lectures as a student at the Swiss Federal Polytechnic School, making it difficult to obtain recommendations for graduate school and teaching positions. Nevertheless, even as a young high school student, Einstein taught himself geometry and calculus from books, and dug into impenetrable books by philosophers Kant and Schopenhauer.[15] Impudence and unruliness are not sufficient to make a genius.

A second myth that needs to be refuted is that geniuses find it easy to solve problems. A student who can't solve a math problem within two minutes typically responds by throwing up his hands, saying, "What do you want from me? I'm not Einstein." However, even a certified genius like Einstein did not sit down one morning and say to himself, "Today, I'm going to invent the theory of relativity," and lo and behold, the theory is ready in time for lunch. In fact, at that time he was not even considered to be a genius, neither by himself, nor by anyone else.

For years, Einstein had thought about an inherent contradiction between Newtonian mechanics and Maxwell's theory of electromagnetism. Many physicists (Hendrik Lorentz, Henri Poincaré (1854–1912) and Max Abraham) had proposed theories in an attempt to resolve this contradiction, but they were unsatisfactory, at least to Einstein. Einstein worked intensively on this problem, but in the spring of 1905, after more than a year or more of unsuccessful attempts to find a solution, Einstein reviewed his progress with his friend Michele Besso (1873–1955), and decided that he would abandon the attempt. By the next morning, he had the solution in the special theory of relativity![16] Later on, Einstein worked for years on an attempt to extend the special theory of relativity to a general theory, which would take acceleration into account. At one point he even published incorrect results that he had to retract.

So what made Einstein a genius, rather than just another run-of-the-mill intelligent person? If you had to choose one trait, it would be intellectual courage, the willingness—in the words of *Star Trek*—to go where no man had ever gone before, together with the intellectual integrity and the perseverance to work out the mathematical implications of his theories in full detail.

The Special Theory of Relativity

The theory of special relativity grew out of two observations. The first was the well-known principle that the laws of mechanics are the same for all inertial observers (those in nonaccelerated motion). There is no experiment that you can perform within an airplane that enables you to discover whether you are stationary on the ground or flying at a cruising speed of more than 10 kilometers/minute relative to the Earth. That is why the flight attendant can calmly pour a cup of coffee, because the relative motion between the coffee pot and the cup is zero. Einstein simply took as an axiom that *all* laws of physics, not just the laws of mechanics, were the same for inertial observers. The second observation concerned the speed of light. Maxwell's theory of electromagnetism predicts a specific speed for light waves (page 27). But this is an absolute speed, with no reference to speed at which the source is traveling. Compare this with the coffee pot, traveling kilometers/minute *relative* to the Earth and at the same time is at rest *relative* to the airplane. Einstein took as an axiom that the speed of light is constant for all observers, and had the intellectual courage to work out the consequences in great physical and mathematical detail without flinching at the weirdness of the results. His main tool was the thought-experiment, in which he posed novel questions about clocks, spaceships, and light waves.

General relativity extended the special theory to accelerated motion. The basic idea came to Einstein one day at his desk in the Bern patent office. He noted that when he tipped back his chair, it seemed to be weightless. Einstein inferred that gravitation and acceleration were indistinguishable. He followed this up with a famous thought experiment, in which he showed that a person in a closed elevator cannot perform an experiment to decide if she is being pulled downward by the force of gravity, or if the elevator is being pulled upward in free space at an acceleration equal to that produced by the force of gravity. The assumed equivalence between gravitation and acceleration implies many strange predictions (such as light waves bent by gravitation), but again, Einstein had the intellectual courage to follow through, though the mathematical difficulties prevented him for many years from producing a satisfactory theory.

The problem that Einstein really wanted to solve—an explanation of the quantum nature of matter and energy—eluded him completely. Max Planck had shown that energy is emitted and absorbed from matter as if it existed only in discrete units called quanta, and later, Einstein himself had extended the idea to show that that light itself was quantized. Despite years of work, he was never able to explain why this should be so, and it was left to other physicists to work out the quantum structure of the atom. If there is an explanation for his failure to produce a theory of quantum mechanics, it is possibly that his courage failed him here, because every attempt led to weird results that even Einstein could not accept, though today they form a major subject in the undergraduate physics curriculum.[17]

Einstein lived hand-to-mouth throughout his studies and for many years afterward, until finally in 1909 he received an academic appointment at the University of Zurich. He was an accomplished violinist and quite sociable. Unlike Newton, Einstein was a "ladies' man," marrying and divorcing twice as well as carrying on affairs. His first wife Mileva Marić Einstein (1875–1948) also studied physics and mathematics, but eventually Einstein cut her out of participation in his research, leaving her to raise the children like a traditional housewife. Mileva (or her lawyer) was sufficiently vigilant to write a clause in their divorce agreement stipulating that should he should win the Nobel Prize, the money would go to her; and so it did when he won the prize in 1921.

5 Pseudoscience: What Some People Do Isn't Science

In chapter 2 we defined a scientific theory and showed that there are claims about the universe that are not scientific. Scientists certainly recognize the existence of fields of human activity like art or religion that are not considered to be within the province of science, and these fields do not necessarily conflict with science. Far more insidious to science and society is the widespread phenomenon called *pseudoscience*. Pseudoscience involves the use of the style and trappings of science for claims that cannot by any stretch of the imagination be called science. In this chapter, I would like to analyze some of the more popular pseudosciences, in order to show why they are not part of the body of science.

Pseudoscience as the precursor of science

Many pseudosciences are frozen artifacts of earlier periods. Before the advent of modern science they were considered scientific, and they retain their ancient doctrines unchanged and unaffected by the achievements of the past few centuries. It is as if followers of pseudosciences would prefer a noisy, expensive, low-precision, heavy, mechanical calculator from the early twentieth century to our quiet, cheap, high-precision, pocket-sized, electronic calculators.

Alchemy was the progenitor of chemistry, as astrology was of astronomy. The benign nature of highly diluted homeopathic concoctions did far less damage than did the medical practices of the eighteenth century like blood letting and purging. Furthermore, many of science's greatest heroes engaged in what is today called pseudoscience. Newton spent years working on alchemy, and Galileo cast horoscopes, including one cast on January 16, 1609, predicting long life for Ferdinand I de Medici, the Grand Duke of Tuscany. Proponents of astrology, especially those who like to bask in the scientific prestige of Galileo, will be sorry to learn that Ferdinand died just twenty-two days later![1]

One of the characteristics of pseudoscience is the almost religious-like canonization of the writings of the "elders." As one nearer the age of the "elders" than the "youngsters," I can appreciate that with experience comes a modicum of wisdom that you wish you had decades ago. But there is no reason to believe that today we are any less intelligent and competent than people who lived hundreds or thousands of years ago. On the contrary, while we respect and admire the achievements of the pioneers of science, the accumulation of knowledge over the centuries gives us a better perspective in which to view the world.

A scientific theory is *never* justified solely, or even primarily, by appeal to authority. At most, authority is deferred to for convenience: since science is so extensive that you cannot be familiar with all the details of all subjects, you have to accept the word of authority concerning outside your specialty. Nevertheless, there is quite a lot of similarity in the way science is practiced across its specialities, so that you can usually judge which authorities are reputable and how to resolve conflicts between them. But under no circumstances is a scientific theory ultimately judged by the stature of its creator or supporters. If a young student doesn't accept Einstein's theory of relativity, she is not told to shut up and refrain from desecrating the memory of one of the most revered scientists of all time; instead, she is quietly shown the direction to the library and invited to read and reread his papers, to check and recheck his calculations and arguments, and to examine and reexamine the experiments that provide confirmation for the theories.

Science and its history

Before continuing the discussion of pseudoscience, let us further analyze the relation of science to its history.

One of the central debates concerning the teaching of science has to do with the place of the history of science in the curriculum. Should science be taught *ahistorically*? That is, should science be taught by simply presenting theories, observations, and experiments, without regard to the historical process of discovery, rich in colorful characters, mistakes, and controversy? Physics can be taught by simply presenting the various theories of gravitation, thermodynamics, and electromagnetism without ever mentioning the names of Isaac Newton, Sadi Carnot (1796–1832), and

James Clerk Maxwell. Darwin's name could be totally forgotten without doing any damage to biology. Many people oppose this approach, because science is part of the human story and should be told as such. Furthermore, if students are shown the challenges facing these pioneers and their struggles to create modern science, perhaps they can put their own difficulties in perspective and persevere in their studies. Regardless of whether you accept or reject the idea that history is an essential part of science education and even of science itself, it seems clear that scientific *results* are universal and hence ahistorical. If advanced extraterrestrials exist and if contact is ever made with them, we have reason to expect that they would know about gravitation and relativity and quantum mechanics, even if they had never heard of Newton and Einstein and Schrödinger.

The ahistorical nature of the results of science, as opposed to the reverence accorded the writings of the pioneers of pseudoscience, is clear when you realize that the actual writings of the pioneers of science are today totally ignored, except by historians and by a few scientists and educators with an interest in history. Newton's *Principia* is simply not accessible to today's scientists. If you insist on reading the original, there is a book that is in effect an exegesis of the original text.[2] The reason that the *Principia* and other works of the period on physics are inaccessible is that they are couched in the mathematics of classical geometry. For example, Galileo provides an elaborate geometrical argument that the integral of x is $x^2/2$, which is a result of elementary calculus that is taught in high school these days.[3] Calculus, the mathematical innovation of Newton and Leibnitz, took quite a long time to develop and was only given a good theoretical foundation in the late nineteenth century.

The dynamic nature of science is clearly demonstrated by the fact that Darwin's extensive written works are outdated and are not used in routine scientific work. Despite the fact that extensive research in the life sciences is carried out at my institution, a search of the online catalog for a copy of *On the Origin of Species by Means of Natural Selection* sent me, not to the well-stocked and bustling life-sciences library, but to the musty historical collection, tucked away in a seldom-visited library. I found and borrowed a one-hundred-year old copy of the *Origin*; reading it through, I was more than once forced to cut through the uncut edges of the pages, which had not been consulted since its publication! Of course, there is no reason for a practicing biologist to consult the *Origin* as part of his routine scientific work, though he may want to do so out of an interest in the historical foundations of his field. The basics of Darwin's theory are accepted as fact and there is no need to work through hundreds of pages of his argument, in particular, Darwin's fascination with artificial selection as practiced by farmers and pigeon enthusiasts. The *Origin* is not used as a textbook in evolutionary biology, because essential aspects of the theory like genetics and molecular biology did not exist at that time, and the book contains many incorrect attempts at answering problems whose solution came only later.

A comparison of *Origin* with a modern textbook on evolution is instructive.[4] The latter covers the basics of evolution fairly quickly and even the fossil record—the target of incessant attacks by creationists—is not surveyed in great depth. While fossils are extremely important in that they provide the raw material for the reconstruction of the phylogeny of life, the details are left to professional paleontologists, because the evidence that fossils provide for the theory of evolution is uncontroversial. On the other hand, chapter after chapter is devoted to the mathematics of population genetics, controversies about the principles of the classification of species, and explanations of the molecular basis of genetics. Darwin may have pioneered the subject of evolution, but to call it "Darwinism" is today a misnomer, because evolution is a dynamic science that has expanded so far beyond what Darwin could have imagined that he would fail a modern examination in the subject without years of further study.

So when you are told that the wisdom of the ancient Greeks proves the truth of astrology, or that alchemy must be true since it was practiced by a great scientist like Isaac Newton, or that you must consult the pioneering opus on homeopathy written two centuries ago by Samuel Hahnemann (1755–1843), you are justified in concluding that you are dealing with a pseudoscience. Science is dynamic as new experimental techniques are developed, new theories are proposed, and new connections are found between disciplines. Some subspecialties become unproductive and uninteresting, while new ones command interest and inspire scientists to commit time and resources to their study. Together with new notations and pedagogical techniques, these changes are reflected in an unending stream of new textbooks.

Over the line of demarcation

Suppose now that a pseudoscience drops its canon and saints, and attempts to present itself as a dynamic, modern science. How can we distinguish it from science? The answer is that the principles of a science must conform to the definition of a scientific theory, above all, the requirement that the theory explain and predict phenomena. This means that the cumulative corpus of observations and experiments must confirm the theory, not falsify it. Even if occasional experiments support a pseudoscientific claim, they can be considered as coincidences that are likely to emerge randomly. Empirical evidence must be *cumulative, continuing*, and *unambiguous*.

It is often claimed that such experiments are not performed or are suppressed because of prejudice or a conspiracy among scientists or others. Does anyone really believe that the granting of a PhD degree in science is conditional on a mafia-like blood oath, in which you promise not to perform research on a list of pseudosciences specified by a godfather? (If there was such a ceremony, I must have missed it.) Everyone knows that a successful conspiracy should be limited to a handful of people; it could never successfully encompass the thousands of new students granted PhD degrees in the sciences every year. Presumably, such a conspiracy would have to be enforced by blackmail and murder, and we would be regularly treated to sensational headlines (sidebar).

Conspiracies in Science

CalTech Researcher Found Shot Ten Times
Lured to an abandoned warehouse in LA
FBI suspects she was using CalTech computers
for astrological calculations
and was close to a revolutionary breakthrough
Director of Research at CalTech Held for Questioning

In fact, the scientific community is quite diverse and unfettered. If there were the slightest reason to believe that there is some scientific validity to astrology, some graduate student or faculty member somewhere would be more than eager to achieve fame, and perhaps even a Nobel Prize, by performing research on the subject. Even funding need not be a problem, although it is claimed that "alternative" therapies are never tested because giant pharmaceutical companies are worried that such remedies will bite into profits on patentable drugs. Annual sales of homeopathic remedies are in the hundreds of millions of dollars (though they are still a fraction of the sales of conventional remedies).[5] Scientific proof of the efficacy of home-opathic remedies would cause their sales to sail through the roof, which is ample justification for companies manufacturing the remedies and venture capitalists to invest in research grants to pay for the labs and graduate fellowships necessary to carry out such a research program.

So why are scientists so obtuse as to reject lines of research arising from pseudoscience? The provocative philosopher of science Paul Feyerabend (1924–1994) wrote angrily about a large group of scientists who signed a petition against astrology, although it was clear that they knew nothing about it. The reason for the reluctance of most scientists to investigate pseudosciences is not prejudice or a conspiracy, but the absence of the other elements that characterize a scientific theory—conciseness, coherence, and mechanism. A serious scientific research program often requires years or decades to carry out, so before committing yourself to such an undertaking, you have to have not just enthusiasm, but also some reasonable expectation of success.

We will now examine astrology and homeopathy to see if they are consistent with our definition of a scientific theory.

Why astrology is a pseudoscience

What exactly does the "theory" of astrology consist of? The basic claim is that the positions of the "planets" within the "constellations" at the time of your birth determine, or at least strongly affect, your nature and your personality, as well as the events of your future life. The first problem with a putative theory of astrology is that it cannot be written on a sweatshirt. For each planet, each constellation, and each possible relationship between them, there is a different "law." No other laws in the universe work this way. In physics, gravitation, quantum mechanics, and relativity operate uniformly on all objects of the universe. In fact, Einstein's amazing achievement was to reduce the number of primitive concepts by identifying mass with energy and gravity with acceleration. In biology, it has been established that the entire genetic code in DNA is composed of only four different molecules called nucleotides, and their arrangement controls the synthesis of only twenty different amino acids from which all proteins are constructed, so biology too can be written on a sweatshirt. Calculating the motion of a particle or specifying the biological function of a protein may be exceedingly complex, but the theories themselves are concise. If astrologers really want their claims to be considered as science, they will have to come up with something that would concisely explain what makes for "Jupiter-ness" or "Gemini-ness."

More serious, however, is the lack of coherence in astrology. I purposely put the words "planet" and "constellation" in quotation marks above, because these are purely conventional. The word planet means "wanderer," because these apparently star-like objects were observed to move relative to the immense number of other stars, which just rose and set together. Once upon a time, it was believed that there were only five such planets— Mercury, Venus, Mars, Jupiter, and Saturn—because only these can be discerned by the unaided eye. Unfortunately for astrology, three more planets have since been discovered: Uranus, Neptune, and Pluto.

Now, to be consistent, either every horoscope prior to the discovery of these planets was in error because it did not take into account the influence of the planets, or, the three new planets are irrelevant and should be ignored. But astrologers can neither disparage their ancestors by insisting that the new planets are important, nor can they supply a reason why Uranus, Neptune, and Pluto should be ignored, while Saturn continues to influence us.

In fact, the situation is even more problematic. Shouldn't the hundreds of thousands of asteroids be taken into account? The asteroid Ceres, at 933 kilometers in diameter, is over one-third the size of Pluto and since it is relatively near the Earth, its influence should be many times stronger that of Pluto. Pluto itself is now known to be one of a group of planet-like objects called the *Kuiper belt*, though almost nothing is known about individual planets of the group. When a future spacecraft does identify them, will astrology have to be reset again?

Since writing the previous sentence, a Kuiper belt object called *Quaoar* has been identified. Quaoar is one-half the size of Pluto and thus presumably influences our destiny, though it will take years to analyze its orbit. Of course for a scientist, this ambivalence does not exist, because "planet" is simply a term used for convenience. You may argue whether Quaoar is a planet or not, but for a scientist, what counts is not the specific term used, but the scientific characteristics of the object: its mass, orbit, and composition. It does not matter one bit whether any particular object is called a planet or a moon or an asteroid or a Kuiper-belt object, and there is not a single aspect of astronomy or astrophysics that depends on "Jupiter-ness" or "Quaoar-ness."

As for the constellations, they have *no* physical meaning whatsoever. A constellation is simply a two-dimensional projection of a subset of the bright stars in a three-dimensional sector of the sky. Individual stars in a constellation may be thousands or tens of thousands of light years distant from each other, and they may be of widely differing sizes and temperatures, because a small, weakly radiating star may appear bright and thus significant simply because it is relatively nearer.

Consider for example, the constellation *Aries*. The top photograph on the next page shows the constellation, but you would have a hard time picking out the stars that compose it unless someone points them out as on the bottom photograph.[6] (Musca Borealis, the "Northern Fly," refers to an obsolete constellation that no longer exists in the IAU list [see below].) With a lot of imagination, you might be able to see a Ram in the four stars. But consider the distances of these stars from us: Mesarthim is 204 light years away, Sheratan is 60 light years away, Hamal is 66 light years away, and 41 Ari is 160 light years away. From a vantage point in the universe "off to the right" of our position, you might still be able to see Sheratan and Hamal, but the other two stars would be out of your field of view. Ascribing "Ram-ness" to these four unrelated stars is totally arbitrary and meaningless, unless you believe that the Earth is a privileged vantage point. But that puts you back into the pre-Copernican dark ages.

The number of constellations and their boundaries is totally arbitrary. The arbitrariness is reinforced when we note that people of other civilizations (for example, the Chinese) saw a different number of constellations and gave them entirely different forms and meanings.[7] Scientists use an official list of 88 constellations adopted in 1930 by the International Astronomical Union (IAU). The twelve constellations of the Zodiac are of somewhat more interest than others, because the Sun and the planets appear to pass through them, though, of course, the Sun and planets are much, much nearer the Earth than the stars forming a constellation. Astrologers should perhaps explain what makes *Aquarius*, *Gemini*, *Leo*, and *Taurus* so special, and why they neglect their cousins *Boötes*, *Camelopardalis*, *Fornax*, and *Puppis*. ("Hi! I'm a *Camelopardalis*. What sign are you?")

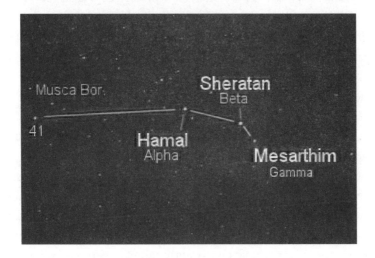

The definition of a constellation is based upon ancient observations performed with the unaided eye; now that telescopes have been invented, the projection of the region of space attributed to a constellation will contain hundreds or thousands of other stars that could not have been observed before its invention. There is no a priori reason to assume that these stars have less influence on our lives than the ones that are interpreted as forming the constellations simply because they were easy for the Greeks and Babylonians to see.

Since the stars are moving with tremendous velocities, the two-dimensional projection changes over time, so the constellations are not an "eternal" characteristic of the universe. In fact, because of a wobble of the Earth's axis called *precession*, the relation of the constellations of the Zodiac to the Earth is continuously changing, completing a full revolution in about 26,000 years. So roughly every $26,000 / 12 = 2,167$ years, your sign moves by one constellation. Astrologers cannot decide whether you should use the traditional signs or the ones that actually appear in the sky on the traditional dates of the year.

A further difficulty is that according to the IAU boundaries, there are fourteen constellations in the Zodiac, invalidating (or not) traditional astrology based upon twelve constellations.

The most basic and central concepts of astrology turn out to be a remnant of the Earth-centered worldview that was demolished by Copernicus, Kepler, and Galileo hundreds of years ago, yet the "theory" of astrology does not take this new knowledge into account.

Can astrology explain and predict, accurately and precisely? Astrological predictions are notoriously vague and lack precision. As for accuracy, while any aficionado of astrology will be happy to provide you with countless instances of anecdotal evidence, empirical studies of the accuracy of astrological predictions have consistently failed to demonstrate any statistically significant accuracy.[8]

The lack of conciseness and coherence and the inability to explain and predict feed off each other. If astrology were in fact a science, then the question of the influence of the outer planets or of the Earth's precession would be decidable from empirical evidence. Simply take large samples of people, cast their horoscopes with and without each of the factors, perform a statistical analysis of the result and obtain the answers. Over a relatively short period of time, perhaps ten or twenty years, the preponderance of evidence would lead to a conclusion that would be accepted by the entire community (or at least by most of its members). Astrology cannot be a science because it lacks the dynamics of a science: theoretical proposals and experimental results leading to a (perhaps temporary) consensus of what is known and what is not. Instead, each astrologist decides what techniques and assumptions to use, which clearly shows that astrology is a belief system like a religion, not a science.

The lack of mechanism in astrology

One of the most important reasons why scientists reject pseudosciences is that they lack even the semblance of mechanism. For astrology to be a scientific theory, there must be some mechanism that explains the effects of the planets and the constellations on the newborn baby. The mechanism must be sufficiently powerful and long lasting so they permanently affect the individual's nature and personality, and influence events decades later. This mechanism could be based upon one of the known forces in the universe, or perhaps there might be an unknown force. Let us first examine the known forces in the universe.

There are only four known forces in nature. Two, the force of gravity and the electromagnetic force, are familiar to all of us. The other two are simply called the strong and weak forces; these forces function within atomic nuclei, binding together neutrons and protons, as well as binding the subparticles from which they are formed. Interactions with these forces cannot take place at distances greater than an atomic nucleus, so they can be ignored by the majority of humankind that does not engage in elementary particle physics. Let us consider in turn if either the electromagnetic force or the force of gravity could provide a mechanism for astrology.

Electromagnetism can be discounted as a mechanism for astrology. It is true that this force is probably the most significant force in our lives because it is the mechanism of chemical interactions, including biochemical interactions. We know, for example, that exposure to x-rays can have serious biological effects, and even the microwave radiation from a cell phone has a biological effect, though there is controversy over whether the effects are clinically significant.

Magnetic fields have much less biological effect, so little in fact, that *magnetic resonance imagery (MRI)* scans are considered to be extremely safe, even though massive magnetic fields are employed. The field of an MRI scanner can be as high as 2 Tesla, which is 40,000 times the strength of the Earth's magnetic field that measures 0.00005 Tesla! In turn, by the time they reach us from space, the magnetic fields of the stars are minute compared to the field of the Earth. Any dangers inherent in an MRI scan come not from the magnetic field itself, but from the effect of the field on metallic devices. For example, a boy was killed when a steel oxygen tank was accelerated into the scanner, and others have been killed or injured due to the effect of the magnetic field on pacemakers and other implanted devices.[9] If astrology claimed that the locations of MRI machines in a hospital determine a newborn baby's personality, the claim might be worth investigating, though all evidence supports the view that even strong magnetic fields have little effect on the body.

Electromagnetic force is transmitted by photons, which travel at the speed of light. There is simply no way that photons sent hundreds or thousands of years ago from the various stars composing a constellation could effect a baby. This radiation is so weak that large telescopes have to be built to detect it and, in a technological tour de force, the Hubble Space Telescope was launched to detect radiation that would otherwise be attenuated or scattered by the Earth's atmosphere. Furthermore, the planets that are so important in astrology do not emit photons, but merely reflect a very small fraction of photons that come from the Sun. If photons could have such an effect, then surely we must consider local sources that bombard the baby with innumerable photons. The baby's astrological chart should take into account the placement of the lights in the delivery room, the type of bulbs, the color of the uniforms of the doctors and nurses, the time of day, and the degree of cloud cover. These should all affect the baby's personality by an enormously greater factor than the positions of the planets.

Consider also that the frame of an ambulance acts as a *Faraday cage*, insulating those inside from the effects magnetic and electronic fields. (The Faraday cage effect explains why frequent strikes of lightening on airplanes do not affect the passengers inside.) If the mechanism of astrology were based upon electromagnetic effects coming from the stars and planets, it would be impossible to cast a horoscope for a person who was born inside an ambulance! Reputable astrologers would have to inform their clients of this complication and inquire into the construction of their places of birth, as a building constructed of reinforced concrete would attenuate the

---

Mesmerism and Magnetism

Magnetism is intensely appealing to pseudoscientists. I suppose that that is because it is the only real, well-known force that appears to be mysterious and can thus be endowed with mystical powers. In the waning years of the ancien régime of prerevolutionary France, a German physician named Franz Anton Mesmer (1734–1815) claimed to be able to cure disease by manipulating "animal magnetism." His treatments became fashionable and trendy, provoking a backlash that led to the establishment of a committee including such luminaries as American diplomat and scientist Benjamin Franklin (1706–1790), pioneering chemist Antoine Lavoisier (1743–1794), and physician Joseph-Ignace Guillotin (1738–1814). Their report from two hundred years ago is a marvelous example of a well-designed scientific investigation; it totally debunked Mesmer's claims.[10] It is surely unfair that Mesmer's name lives on in the verb "mesmerize," while Lavoisier had his head chopped off by a machine named after Dr. Guillotin.

---

electromagnetic influence more than a wooden house. Astrologists simply cannot explain why a few measly photons from stars that cannot even be seen in an urban area have any biological effect whatsoever.

The situation with gravitation is similar. We may find the force of gravity to be overpowering when we have to get out of bed in the morning, but in fact, gravitation is an extremely weak force, so weak that a puny magnetized screwdriver can easily overcome the force of gravity exerted on a screw by the Earth, a gigantic sphere composed of dense rock and metal. Astronomers Roger Culver (1940–) and Philip Ianna have great fun computing and comparing gravitational forces.[11] The force of gravity exerted by the obstetrician is far greater than the sum of the gravitational forces exerted on the baby by all the planets and stars in the universe combined! If gravitation were a significant factor, then presumably a young mother should be told something like: if you want your baby to become a scientist, ask the obstetrician to stand to your left, though if you want him to become a musician, ask her to stand on your right.

So the only possibility that we are left with is that astrological effects are caused by some unknown force. While this position may have been tenable in the time of the Greeks, in our day this is simply nonsense, because the four forces have been studied in great detail and they are able to account for almost all phenomena known in the universe. The remaining "gray areas" concern subatomic particles and the conditions that existed during the first millionth or billionth of a second following the big bang theory. While it is conceivable that an unknown force may exist, it is simply inconceivable that a force exists whose *only* effect is upon the personality and future of humans, and whose effect is determined at the moment of birth.

The assumption that there exists an unknown force with biological implications leads to too many unanswerable questions: Why does it influence a 3 kg baby at the moment of birth and not a one-cell zygote at the moment of conception? (Presumably to avoid asking your parents embarrassing questions and exposing delicate family secrets.) Since all mammals are biologically very similar, is a dog's personality influenced by the stars and planets? (Yes! An Internet search of "dog astrology" yielded almost 100,000 hits.) Well, how about a mouse or a cockroach or a tomato or a bacterium? Precisely the same molecular processes have been shown to occur in all life forms, so why should the unknown force be different. All the discoveries about the universe portray a totally different situation of universal laws that affect all objects identically, whether they are animal, vegetable, or mineral.

They laughed at Galileo

At this point, one expects to hear something like: "They laughed at Galileo, too," meaning that forces were unknown until they were discovered, and that pioneers were laughed at by the stodgy scientific establishment. Presumably, astrology will be vindicated some day. However, this is a total misrepresentation of the history of science. First of all, no one laughed at Galileo; the Inquisition would not have bothered itself persecuting clowns and jesters. Galileo was persecuted precisely because his advocacy of the Copernican system *was* convincing and was considered a threat to the theology and political interests of the Roman Catholic Church at that time. Great scientific discoveries are not necessarily immediately accepted, but the transition period from disbelief to acceptance is relatively short, a few decades at most, as scientists perform experiments and work out implications of a theory, until the preponderance of evidence convinces scientists to accept the theory. Novel ideas are just too interesting to ignore. If there were any reason whatsoever to believe that there is an unknown force that can function as a mechanism for astrology, scientists would compete for the honor of discovering the details.

If there is in fact a fifth force, it will be discovered during some scientific experiment, and it certainly won't fit the shaky edifice of an ancient system, just as the development of modern chemistry grew out of experimentation and theory building and never justified the claims of alchemy. Scientists refuse to study astrology, not because of prejudice or because there is a conspiracy afoot, but simply because there is not a shred of evidence that would justify the expenditure of valuable time from a career.

Could there be a mechanism to explain homeopathy?

Homeopathy was developed by Dr. Samuel Hahnemann (1755–1843). He based his method of treatment upon the *Law of Similars*: Disease can be cured by ingesting extremely small amounts of substances that in large doses cause symptoms similar to the disease in healthy individuals. I won't present a analysis of homeopathy to the depth of the previous analysis of astrology, because the essence of the analysis is the same: Homeopathy does not possess any of the characteristics of a scientific theory except for its outer trappings.[12] However, it is worth discussing one aspect, namely, the mechanism.

In the eighteenth century, there was nothing a priori impossible about the Law of Similars. In the context of the medical knowledge at that time, it was conceivable that during the ensuing decades a mechanism would have come to light. However, precisely the opposite occurred.

From chemistry, the study of molecules and their interactions, we now know that at the incredibly high dilutions used in homeopathic remedies, most pills or elixirs will not contain even a single molecule of the purported active ingredient. Recognizing this, homeopathy dropped its chemical claims in favor of a theory that the water retains a "memory" of the active ingredient. For this to be taken seriously, there must be a plausible mechanism that explains how molecules of water store the memory of the extract of duck liver or whatever it came into contact with before being diluted. The discussion of the impossibility of mechanism in astrology applies here too. There are only four forces known to nature and the existence of a fifth whose effects appear only as required by homeopathy is nonsense. H_2O is a very simple molecule and its physics and chemistry have been well understood for decades. Water is water is water and there is no place for a "memory" to hide in the structure of the molecule.

The impossibility of obtaining a biological effect from extremely diluted solutions becomes apparent when compared with the concentrations of other substances that must be ingested in order to produce a measurable effect. Some people believe that conventional medicines are harmful and avoid them, but a glance at their dosages furnishes a guide as to what amount of a drug can actually cause an effect on a human body. Headache pills contain 500 milligrams of the active ingredient and tranquilizers typically contain from 1–5 milligrams; each milligram contains billions upon billions of the molecules required to cause a clinical effect. Even "native" remedies like the bark of the willow and the cinchona trees were eventually analyzed and found to contain large quantities of real active substances: aspirin and quinine.

Pathogens face the same hurdle. We remain healthy in the presence of the millions and billions of bacteria and viruses we come into incessant contact with in our environment. Only when the exposure passes a certain level can bacteria and viruses overcome the natural and effective immune system of the body.

To get a feel for the numbers, look at the *Foodborne Pathogenic Microorganisms and Natural Toxins Handbook* (affectionately known as the *Bad Bug Book*) published by the Center for Food Safety and Applied Nutrition of the US Food and Drug Administration.[13] For *Vibrio cholerae*, the pathogen that causes the terrible and often fatal disease cholera, an infective dose—the number of bacteria that you have to ingest to become ill—is one million! Additionally, during the incubation period, the bacteria will rapidly reproduce until extremely large numbers are churning out toxins at sufficient concentrations to produce the symptoms. Put another way, you could drink an elixir containing 10,000 cholera bacteria and nothing should happen to you. (To be on the safe side, don't try this at home!) So how can you even conceive of one single molecule curing a disease when it takes a million bacteria to cause it? It just doesn't make sense.

Conversely, it is not the minute amount of weakened bacteria or viruses in a vaccine that protects you, but the massive amount of antibodies produced in reaction by the immune system. These antibodies are detectable in modern lab tests and this mechanism is an essential part of the scientific justification that was not available to Hahnemann's contemporary Edward Jenner (1749–1823) who first performed vaccination for smallpox.

We see in homeopathy the same characteristics that led us to classify astrology as a pseudoscience: the reverence for elders overcoming the lack of predictive success and the refusal to deal with contradictions between the putative underlying explanatory mechanisms and newly discovered knowledge about the basic sciences.

The waste of pseudoscience

To the extent that people pursue pseudosciences as a hobby or diversion, the practice is harmless. Many of the tricks used by pseudoscientists can be mastered in the course of learning to be a magician.[14] Certain pseudoscientific practices can be beneficial if they give people the spiritual comfort, psychological support, or relief from psychosomatic ailments that comes from having someone listen attentively to your problems.

The problem begins when pseudosciences are not satisfied with their status as belief systems and claim a scientific mantle that they do not deserve. As noted long ago by David Hume (1711–1776):

> The knavery and folly of men are such common phenomena, that I should rather believe the most extraordinary events to arise from their concurrence, than admit of so signal a violation of the laws of nature.[15]

Enormous resources are invested in pseudoscience that could be better invested in improving the health and education of the public. Furthermore, the advice given by pseudoscientists frequently causes real damage to those who seek its advice. One only has to think of the tragedies that can occur if one's choice of a mate is dictated by astrological signs; the process is sufficiently unreliable as it is that to deliberately introduce additional randomness is unconscionable.

Finally, this travesty of science is particularly saddening because it seduces talented young people into dedicating their lives to a charade when they could be more satisfactorily employed elsewhere in science, education, or health care (though probably with less remuneration). We can only hope that education will eventually triumph over pseudoscience as Charles Mackay vainly hoped over 170 years ago:

It is to be hoped that the day is not far distant when lawgivers
will teach the people by some more direct means, and prevent
the recurrence of delusions like these, and many worse, which
might be cited, by securing to every child born within their do-
minions an education in accordance with the advancing state
of civilization. If ghosts and witches are not yet altogether ex-
ploded, it is the fault, not so much of the ignorant people, as
of the law and the government that have neglected to enlighten
them.[16]

* * *

Pseudosciences fail to comply with most or all of the characteristics of a
scientific theory. They are rarely concise and coherent, and almost never
expressed in mathematical laws. Pseudoscientists studiously ignore any
discrepancies between their predictions and the real world, attributing such
failures to bad auras emanating from skeptical observers. If a mechanism
is supplied, it is invariably in the form of a mysterious energy field, un-
detectable by any other means. One can only marvel at the audacity of
pseudoscientists who blithely purvey the wildest fictions as science.

LOUIS PASTEUR: SERIAL SCIENTIST

Louis Pasteur (1822–1895) did not have the luck to be born into a well-to-do family like Charles Darwin. He was an extremely talented artist, and it is not farfetched to imagine that if circumstances had been different, he might have become as famous as other nineteenth-century French artists. While not an outstanding student, Pasteur was, nevertheless, sufficiently talented that his teachers predicted that he could become a college professor, and his parents made the effort to send him to the prestigious Ecole Normale Supérieure in Paris. Pasteur studied chemistry and obtained a doctoral degree, performing pioneering studies on crystals. He was able to show that the compound tartaric acid existed in two independent forms, one left-handed and one right-handed.

By 1854, Pasteur was professor of chemistry at the industrial city of Lille in northeastern France. He was asked to look into difficulties in the fermentation of alcohol from sugar. Sometimes, a batch of solution undergoing fermentation produced the desired alcohol and sometimes the result was unpredictably spoiled. In a series of experiments, Pasteur was able to demonstrate that fermentation is a *biological* process, and that alcohol is the waste product of the digestion of sugar by cells of yeast. The spoiled batches were the result of similar action by other microorganisms. The technique he invented for preventing the spoilage of food—heating to destroy microorganisms—is called *pasteurization* in his honor.

We sometimes talk of a serial criminal compelled to repeat his crimes; Louis Pasteur's biography is that of a serial scientist, compelled always to seek out new problems to solve, and his talents led him to a series of pioneering investigations. The foray into fermentation marks the beginning of Pasteur's work in biology, which led him out of his official speciality, chemistry. Studies of the microorganisms associated with fermentation led to experiments refuting the spontaneous generation of life. The study of microorganisms led to the study of disease, first in silkworms, then in farm animals (anthrax), and finally in humans (rabies).

A serendipitous discovery led to the development of modern vaccination by injecting cultures of weakened microorganisms. A forgotten batch of cholera cultures from chicken was found to be incapable of infecting other chickens, but these same chickens did not contract cholera when injected with a fresh culture. The study of rabies showed that disease could be caused by "filterable viruses" that were so small that they could not be viewed under a microscope. It also turned out that—unlike bacteria—viruses could be grown only on organic material, not in sterile chemical solutions. Still Pasteur was so confident of the results of his experiments that he agreed, albeit with some trepidation, to try the first vaccination against rabies in 1885.

In 1857, Pasteur had returned to the Ecole Normale Supérieure in Paris, but in 1868 he was stricken by a stroke that left him semi-paralyzed. Nevertheless, he continued his scientific work until near the end of his life.

Louis Pasteur was blessed with an intuition that led him to scientific results that others could not see, but he refused to accept any idea that was not confirmed by painstaking experimentation. Pseudoscientists, especially those promoting untested health products, would be well advised to heed Pasteur's words:

> This marvelous experimental method, of which one can say, in truth, not that it is sufficient for every purpose, but that it rarely leads astray, and then only for those who do not use it well. ... The charm of our studies, the enchantment of science, is that, everywhere and always, we can give the justification of our principles and the proof of our discoveries.[17]

6   The Sociology of Science: Scientists Do It as a Group

Is reality real?

In a sense, virtual reality—the simulation of reality by computers, sensors, and displays—is more real than reality. This is because you can always step outside a simulation. If you have a question or a conjecture as to how the system works, you can always ask the computer engineers who developed the system to look at their hardware and software in order to give you a definitive answer. When it comes to science that studies the "real" universe, this luxury is denied us. Scientists cannot step outside the universe and see if their theories are correct.

Many people believe in a divine being who created the universe and whose existence transcends the physical universe. Presumably, the divine being could answer our questions about the structure of the universe, but existing religions contain no information whatsoever as to the correctness of relativity or quantum mechanics. Even if these topics appeared in religious doctrines, we would have no way of deciding among the different answers that would almost certainly be given by different religions, just as they give different answers concerning theological questions. Scientific theories would then be accepted upon faith rather than upon evidence.

Science can never claim to have absolute truth about reality, or even to prove absolutely that the universe itself exists. It is always possible, as Shakespeare wrote, that we have been set upon a stage to act out a drama:

> Life's but a walking shadow, a poor player
> That struts and frets his hour upon the stage
> And then is heard no more.
> It is a tale
> Told by an idiot, full of sound and fury,
> Signifying nothing.[1]

Two literary works try to express the distress and confusion of those trapped within a universe but nevertheless afforded a glimpse of what is outside that universe. In 1884, Edwin A. Abbott (1838–1926) wrote a charming book called *Flatland: A Romance of Many Dimensions*. The book is narrated by "Square," a denizen of a two-dimensional world inhabited by two-dimensional geometric figures. He is treated to a glimpse of the third dimension, but, needless to say, Square finds it hard to convince his colleagues of this discovery and eventually loses faith himself, unable to decide if the third dimension—"Upward, not Northward"—is real or just a dream. Jostein Gaarder's (1952–) novel *Sophie's World* is both a popular introduction to philosophy and a clever dramatization of the age-old existential question: Can we know that we are not just characters in a novel or actors in a dream? These philosophical questions are unlikely to be resolved, and science, like other human activities, must function within this state of uncertainty.

In the absence of a source of correct answers, the acceptance or rejection of scientific theories is performed by the scientists themselves. As human beings, scientists are subject to the biological and psychological factors that affect us all. Furthermore, scientists live within a society composed of many cultures, and create their own cultures within the institutions (universities, research labs, and professional societies) in which they function. During the past few decades it has become clear that the social aspects of science are as much a part of its nature as the philosophical aspects, and probably more important in influencing the actual practice of science. In this chapter, we present the pioneering work of Thomas Kuhn on the sociology of science, followed in the next chapter by a discussion of the postmodernist attacks on science that grew out of Kuhn's work.

Communities of scientists

Recall from chapter 4 that Albert Einstein insisted that his theory of special relativity was correct even though Walter Kaufmann claimed that his experiments had falsified the theory. What makes Einstein a great scientist, while a person who makes unscientific claims that are also falsified by experiments is labeled a quack, a charlatan, or worse, a fraud? Clearly, it is *not* because we have independent access to reality. The answer is that Einstein was a full-fledged member of the *community of scientists*, in particular, the community of physicists. He had studied physics at the Swiss Federal Institute of Technology, he wrote papers using the language and notation common to other physicists, he presented his papers at conferences and submitted them to scientific journals, and he was working on problems that were of deep interest to other physicists of his day. His ideas were novel and difficult to understand, and they had revolutionary consequences for physics, but they were within the tradition of research that was being carried out at that time.

The theory of special relativity was not immediately accepted by all other physicists, and it had to compete with other theories that were suggested at the same time, but Einstein had just as much right to insist that his theory was true and that Kaufmann's experiment was in error as Kaufmann had in asserting that his experiment gave the correct results thus refuting the theory. As the consequences of special relativity were worked out and as more experiments were proposed and carried out, this dispute was eventually settled by the accumulated evidence.

An analysis of the social aspect of science was first presented in one of the most famous and controversial books in the philosophy of science: Thomas Kuhn's (1922–1996) *The Structure of Scientific Revolutions* first published in 1962. The progress of science according to Kuhn is shown in the following diagram:

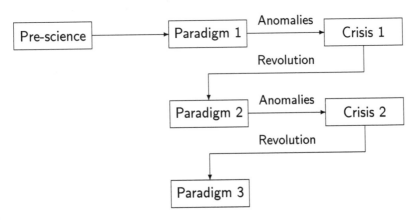

Initially, there is a phase called *pre-science*, in which many ideas are proposed without any one becoming sufficiently convincing to be accepted. Eventually, according to Kuhn, scientists coalesce around a set of concepts and techniques, forming communities whose members share a framework consisting of values as to what constitutes an interesting scientific problem, what techniques are to be used in theoretical and experimental investigations, and what interpretations are reasonable. Kuhn called these frameworks *paradigms*.[2] Scientists working within a paradigm are engaged in *normal science*, solving problems ("puzzles" was Kuhn's somewhat condescending term) and refining techniques over the years.

As the period of normal science continues, despite progress in solving problems, there will remain a set of problems called *anomalies* that obstinately resist solution. Finally, a *scientific revolution* occurs in which entirely new ideas are used to resolve the anomalies. Scientists working with these new ideas form a new community, while those who do not accept the new ideas remain in the old one, which may eventually disappear if the work they do is no longer fruitful.

Kuhn rejected Popper's falsificationism; he claimed that scientists do not undergo revelations in which they suddenly understand that the theory they hold has been falsified and that a new theory better explains the anomalies of the old one. Instead, they often stubbornly hold on to the old theory, which are discarded only as members of the old community die out, while young students join the new community instead.

It is uncontroversial that scientists belong to communities and that these communities are replaced, or at least modified, as time goes on. Modern scientific work demands specialized knowledge and resources that can only be obtained in universities and research institutes set up for that purpose. These institutions are communities that accept aspiring entrants according to their own criteria. Advancement to full membership is attained by publishing articles in journals and presenting papers at conferences, and these journals and conferences run by scientific organizations that form the broader community consisting of all scientists specializing in that field.

While communities rarely die off with their members, it is certainly true that they metamorphose as scientific interests change. New organizations and journals are founded, and old ones change their focus. For example, I belong to a large organization called the *Association for Computing Machinery*, which was established in 1947 just as the first computers were being built. Today the publications of the association rarely mention computing "machinery"; instead, the organization has metamorphosed into one for all computing professionals, and articles in their publications cover modern topics like the Internet.

It is also uncontroversial that the process of science is influenced by the societies in which scientists live. In particular, the allocation of resources is strongly influenced by politics. Throughout the Cold War, the defense budgets of the superpowers accounted for much of the spending on scientific research. For example, the Internet grew out of research in a communications technology called *packet switching* that was sponsored by the US Department of Defense, which was interested in the construction of reliable communications networks.

Today we are witness to the continuous political pressure that is applied in order to influence the allocation of resources to medical research. Women's groups press for research on breast cancer, while men's groups lobby for research on prostate cancer. There has been massive investment in research on vaccines and treatments for AIDS, despite the fact that—at least until recently—AIDS killed far fewer people each year than malaria and tuberculosis. Research on these diseases has been funded at relatively low levels, possibly because they are rare in developed countries that have sufficient resources to carry out medical research.

Incommensurable paradigms

Kuhn's book is controversial because he stated that communities "practice their trades in different worlds," or, in a well-known term, that paradigms are *incommensurable*.[3] The idea is that revolutionary theories like evolution, relativity, or quantum mechanics create new communities, whose members work on problems and use language, concepts, and notation in a manner that is so foreign to the existing communities that the members of the two communities cannot fully understand each other's work: "a law that cannot even be demonstrated to one group of scientists may occasionally seem intuitively obvious to another."[4] Furthermore, since the standards used by the communities may be different, they may not be able compare their results with each other:

> Some old problems may be relegated to another science or declared entirely "unscientific." Others that were previously nonexistent or trivial may, with a new paradigm, become the very archetypes of significant scientific achievement. As the problems change, so, often, does the standard that distinguishes a real scientific solution from a mere metaphysical speculation, word game, or mathematical play.[5]

The communities become estranged from each other and the new ones triumph not by the intellectual conversion of the members of the old community as they weigh the evidence that supports new theories, but simply by natural replacement as the old community dies out.

It is true that the problems that are at the forefront of scientific research may change even in very significant ways. For example, in the nineteenth century, the best chemists devoted their careers to measuring the atomic weights of elements, and there was much debate as to whose measurements were the most accurate. With the advent of nuclear physics in the twentieth century, it turned out that the chemical measurement of atomic weights is a meaningless exercise, because any naturally occurring sample of an element is a mixture of various isotopes with different atomic weights. In effect, the problem was declared to be "unscientific." The question we need to ask is not if science changes, but if Kuhn's model of paradigms, revolutions, and communities faithfully represents the historical record.

In the following sections, I would first like to claim that the historical record does not support Kuhn's claim of incommensurability, and then discuss in detail the philosophical and social implications of the claim. Here is a list of the major candidates for scientific revolutions where you would expect that the paradigms would be incommensurable: (a) the transition from an Earth-centered universe governed by the distinct natures of terrestrial and heavenly matter to a Sun-centered solar system governed by mathematical laws of motion and gravitation; (b) the transition from this formulation of classical physics to the modern physics of relativity and quantum mechanics; and (c) the transition from the pre-science of the special creation of species and from Lamarck's theory that acquired characteristics are inheritable to the theory of evolution by natural selection. The controversial case of cold fusion is then presented as an example of two incommensurable paradigms, though one of the communities will probably turn out to be pseudoscientific rather than scientific.

Classical physics: a new community?

Nicolaus Copernicus never lived to see his work *On the Revolutions of the Celestial Bodies* published. Its posthumous publication was accompanied by a preface written by his disciple Andreas Osiander (1498–1552) disclaiming any physical meaning to Copernicus's work and presenting it merely as a mathematical convenience. Therefore, the work was not perceived at the time as being revolutionary and languished for many years. Kepler's three laws were the first correct description of planetary motion and were based on the Copernican Sun-centered universe, and while Kepler suffered from the vicissitudes of the Thirty Years' War (1618–1648) in central Europe, he was not persecuted for his scientific writings. After Newton published his *Principia* in 1687, his work was immediately and enthusiastically accepted by the entire scientific community: Newton was knighted and granted awards and prestigious positions by the English government. He did not have to deal with an existing community that could not understand or utilize his findings; if anything, he had to deal with disputed claims of priority by Robert Hooke and Gottfried Wilhelm Leibnitz, and Newton engaged in bitter feuds with them and their supporters.

So if there was a point at which a Kuhnian revolution took place, it must be centered on Galileo and his trial by the Sacred Congregation of the Holy Office of the Roman Catholic Church (also known as the Inquisition).

Galileo is credited with many scientific achievements, the most revolutionary of which is the mathematical analysis of motion, in particular the formulation of the principle of inertia and the correct definition of acceleration. But he was not put on trial for these, nor does there seem to be any indication that distinct communities formed, either for or against his theories. It is even incorrect to speak of a new paradigm in terms of novel interests and techniques, because Galileo continued to use classical geometry to study classical problems of motion.

Galileo was tried for his propaganda (to use Feyerabend's term) in favor of the Copernican system of a Sun-centered solar system. Galileo's writings on the Copernican system *were* understood by the existing communities of astronomers and mathematicians, and that is why he was put on trial: the system was seen to be a threat to the theology of the Catholic Church and (perhaps more important) to its political power.[6]

Galileo maintained close connections with Church scientists, and some influential members of the clergy were initially sympathetic to the Copernican system. But Galileo was not a tactful politician versed in the arts of diplomatic evasion and understatement. Had he been a bit less abrasive and more willing to compromise on the public presentation of his views, he might have escaped the trial and subsequent house imprisonment that lasted the rest of his life. Thus, although it is easy to see the work of

Copernicus, Kepler, Galileo, and Newton as a scientific revolution (more properly as the transition from pre-science to a scientific paradigm), it is more difficult to interpret it within Kuhn's framework of incommensurable paradigms.

Modern physics: a new community?

In terms of new concepts and techniques, no transition in physics, or even in all of science, is comparable in its revolutionary aspects to the abandonment of Newtonian mechanics in the first half of the twentieth century in favor of the theories of relativity and quantum mechanics with their nonintuitive predictions like time dilation and the probabilistic behavior of particle waves. Yet there is no evidence for the creation of communities that could not communicate with each other. Certainly, first-rate scientists like Henri Poincaré and Ernst Mach never accepted Einstein's theory of relativity, and many other scientists did not understand it initially. (An anecdote is told about Cambridge astronomer Arthur Eddington who, upon being congratulated as one of the three people who understood relativity, replied that he could not imagine who the third could be.)[7] But Einstein's paper was published in a leading German physics journal *Annalen der Physik* and relativity was subjected to the normal heated debate at scientific conferences. Hendrik Lorentz never accepted relativity, but how can we say that he and Einstein inhabited different communities when Lorentz's theory contained some ideas similar to Einstein's, and it was Lorentz's equations that Einstein adapted for his theory of relativity?

When it came to quantum mechanics, Albert Einstein found himself on the nonaccepting side of the controversy. His ambition to solve the problems of the structure of matter was far greater than his ambition to solve the problems of space and time for which he is famous, and he was accomplished in the techniques of statistical mechanics that seemed to hold the key to solving these problems. Yet Einstein never accepted quantum mechanics, because he refused to accept its nondeterministic nature, saying that "God does not play dice with the universe." But again, there is no evidence that Einstein could not understand quantum mechanics, appreciate its results, or utilize its methods should he be so inclined. On the contrary, he was one of the formulators of the Einstein-Podolsky-Rosen thought experiment, which used the techniques of quantum mechanics to argue that it was incompatible with special relativity. Experiments carried out by Alain Aspect (1947–) in 1982 showed that the phenomenon described by Einstein, Boris Podolsky (1896–1966), and Nathan Rosen (1909–1995) does in fact occur, and that it is just one more of the weird results of quantum mechanics.

Kuhn claimed that criteria for comparing theories can be so different that comparison is impossible. Einstein *could* compare quantum mechanics with classical mechanics; he just *chose* not to accept the results. Scientific disagreements do not translate into incommensurable paradigms.

Evolution: a new community?

Charles Darwin's theory of evolution by natural selection was one of many similar theories on the origin of species that were propounded during that period by others, such as Jean-Baptiste Lamarck and Erasmus Darwin (1731–1802), Charles's grandfather. Darwin's theory was immediately understood, the evidence was discussed and weighed, and the theory was quickly taken up by scientists. In a relatively short period of time, examinations at Cambridge University went from asking students to justify the immutability of species to asking them about the evolution of species.[8]

Obviously, the theory of evolution can be interpreted as having theological implications, and was bitterly attacked by scientists, clergy, and laypersons. The incommensurability was between the *pre-science* of special creation, and the *science* of the theory of evolution. There is no evidence of incommensurability *within* the scientific community, in the sense of scientists not understanding each other or not communicating with each other or holding to different standards of correctness. As in the case of Galileo, fierce opposition to Darwin's theory arose precisely because the theory and the evidence for it were all too well understood and perceived as a threat to the established religion. To claim that the current conflict between evolutionary biology and creationism is a case of incommensurable paradigms of science is to accept religion as science. Of course this is the aim of creationists, but Thomas Kuhn, himself a physicist, would never have agreed to such an extension of his analysis of science.

Cold fusion: a new community?

A few years ago, the scientific world saw either the creation of a new community and incommensurable paradigm just as Kuhn claimed, or the birth of a new pseudoscience. In 1989, two scientists at the University of Utah, Stanley Pons (1943–) and Martin Fleishmann (1927–), claimed to have performed a relatively simple laboratory experiment that produced energy from the fusion of nuclei of deuterium (an isotope of hydrogen with a neutron in addition to the proton in its nucleus).[9] Fusion normally occurs only under conditions of extreme temperature or pressure that occur in stars or hydrogen bombs. It is possible to demonstrate fusion in hot plasma confined within massive magnetic fields, but all attempts at long-term containment of the plasma have failed. Since controlled fusion would radically change the way the world obtains its electrical energy, a frenzy of hysterical media reports erupted, because Pons and Fleishmann claimed to have produced energy from a cup of heavy water and a couple of metallic bars made of platinum and palladium. The phenomenon was called *cold fusion*, to distinguish it from fusion produced in hot plasma.

Now there is nothing a priori impossible about cold fusion; weird things are known to happen in the quantum world of elementary particles and this would be just another one of them, albeit a phenomenon of extraordinary interest. The problems started when other scientists were not able to replicate the experiment and began to question the experimental techniques of Pons and Fleishmann. Furthermore, theoreticians showed that the basis of cold fusion, the catalization by palladium of the nuclear reaction, is impossible.[10] Even the supporters of cold fusion agreed that evidence of fusion, the release of heat energy or beams of neutrons, does not reliably occur in experiments. The US Department of Energy reviewed the claims and the evidence, and made the reasonable recommendation that basic research continue, but the concluded that there was insufficient evidence to warrant a crash program to develop energy resources based upon cold fusion.[11]

In normal circumstances, this controversy, like other scientific controversies, would have simmered in journals and conferences, eventually to be resolved when the preponderance of evidence supported one side or the other. Unfortunately, the stakes involved, both financial and in terms of prestige, were so high that two separate communities developed: the mainstream scientific community, which claims that there is no real evidence that cold fusion occurs, and that proponents of cold fusion are at best being misled by wishful thinking like Blondlot; and the cold fusion community, which claims that the results are real and that they are being unfairly portrayed as pseudoscientists by the mainstream community, eager to protect its research programs based upon hot plasma.

In retrospect, Pons and Fleishmann are ultimately responsible for this mess, because they committed a grave breach of scientific etiquette: they published their results at a press conference instead of submitting them to a scientific journal or conference. If they had done the latter, the controversy would have played itself out within the communities of chemists and physicists, taking perhaps a decade or so to resolve. Depending on the results, either Pons and Fleishmann would have duly received the Nobel Prize (together with Steven Jones [1949–] who had obtained similar results), or, their papers would have been relegated to dusty stacks in scientific libraries. The short-circuiting of the normal scientific process in favor of an announcement to the media led to the founding of a separate community that is probably destined to continue its existence as a marginal pseudoscience.

The social process of science

The assessment of Kuhn's portrayal of science depends to some extent on the definitions of the terms "revolution" and "incommensurable" that are used. While it seems clear that scientific revolutions do occur, and that one scientific paradigm is replaced by another, I have tried to refute the claim of incommensurability in Kuhn's sense of clashing communities of scientists who cannot even compare results because their criteria for judging scientific results and theories are so different. The historical record shows that radically new scientific theories were taken up quickly (within a decade or two), and after a reasonable amount of controversy as one would expect.

Kuhn's book is important not for his historical reconstruction of science, but for his emphasis on the social aspects of the process of science. Theories are proposed and experiments carried out, but before they are accepted as part of the scientific consensus, a long process ensues during which scientists try to understand the theories and evaluate the results of experiments. This process is frequently accompanied by acrimonious debate and many scientists will refuse to abandon existing theories, whether from true conviction or from personal motives like jealously and preserving their own reputations. In most cases, the preponderance of evidence eventually leads to a consensus that almost all scientists accept. In some cases, however, the evidence is not sufficient to convince everyone and competing theories persist, perhaps to be resolved in the future if more evidence arises, or perhaps never to be resolved.

Consensus arises naturally, as there is no authority charged with resolving controversy. Even when consensus is achieved, a controversy can always be reopened if someone comes up with a new theory or new evidence. But as British philosopher David Hume claimed, extraordinary claims demand extraordinary proof (see sidebar), so that it is harder to get people to consider objections to the basics of a well-established theory like quantum mechanics than it is when a speculative theory is concerned.

Hume's Famous Quote

"Extraordinary claims demand extraordinary proof" is a pithy condensation of the following verbose sentence:

Suppose, for instance, that the fact, which the testimony endeavours to establish, partakes of the extraordinary and the marvellous; in that case, the evidence, resulting from the testimony, admits of a diminution, greater or less, in proportion as the fact is more or less unusual.[12]

We have now essentially answered the question: How are results accepted as true by the scientific community? Acceptance comes through a social process of debate carried out in writing through books, journal articles, and conference papers, as well as in face-to-face meetings during lectures, seminars, and meetings at conferences. The process is not always smooth: a well-respected scientist can influence the short-term outcome by withholding invitations to present papers at conferences or by rejecting papers for publication in a journal. As science has passed from the province of amateur gentlemen to the business of well-funded labs, the hand that controls the funding agency has even more power over the scientific progress than ever before.

Nevertheless, it appears that in the long term, the scientific process is *self-correcting*. There is no worldwide guild that controls science, so any group can establish its own organization complete with the necessary paraphernalia of journals and conferences. Theories and research areas are in constant competition for the attention of fresh graduate students and young faculty members ambitiously seeking to make their mark upon science. Science is international, so if one country is in the grip of a flawed theory, scientists in another country can carry forward work on competing theories.

* * *

Our definition of a scientific theory is impersonal: a theory "is," "can be used," "includes." Kuhn was clearly correct in emphasizing that people propose theories, people use theories, and people can accept or reject theories. The study of the social aspects of science has been very fruitful in helping to understand the social processes that have been and continue to be important in the development of science, both in its triumphs and in its failures. Even more important, this study has enabled us to identify and strive to overcome implicit and explicit barriers that have excluded many from participating in science. However, one must be alert not to slide into relativism. Just because people doing science are embedded in a particular social and cultural milieu, it doesn't follow that science is not universal. This issue is discussed in greater detail in the next chapter.

EMMY NOETHER: AGAINST THE CURRENT OF DISCRIMINATION

Emmy Noether (1882–1935) was born to a middle-class family in Erlangen, Germany, where her father was a mathematician. After attending a high school for girls, she became a teacher of French and English, precisely the type of occupation that was expected of middle-class women. Noether decided to attend classes in mathematics at the University of Erlangen, although women were not allowed to formally take courses and sit for examinations. After passing the required matriculation examination, she moved to the University of Göttingen, later returning to Erlangen where she received a PhD degree in 1908.

Noether's research papers attracted the attention of the foremost mathematicians of that day, David Hilbert (1862–1943) and Felix Klein (1849–1925), who in 1915 invited her to join the faculty at Göttingen. Unfortunately, the male chauvinistic practices of that era could not be overcome and she was denied a position, despite Hilbert's caustic remark that: "After all, we are a university and not a bathing establishment."[13] It was not until 1919 that she obtained any sort of official position at the university.

In 1933, the Nazis forced all Jewish scientists including Noether from their academic posts. She emigrated to the United States and began teaching at Bryn Mawr College, where for the first time she encountered women in regular academic positions, including the department head, Anna Pell Wheeler (1883–1966) who had also studied at Göttingen. Sadly, just two years later, Emmy Noether died after an operation.

Noether made immense contributions to abstract algebra, which have been recognized by the ultimate scientific accolade of having her name written with a lower-case character: *noetherian rings*. Hilbert was primarily interested in collaborating with her because of her expertise on invariants, mathematical formulas that do not change when modified in certain ways. She proved an extremely important theorem showing that invariants in the mathematical laws of physics imply that laws of conservation must hold (sidebar).

Emmy Noether's achievements show that discrimination is self-defeating in that it can deprive science in general and specific institutions in particular of the most talented people. She should be an inspiration to all those who have to overcome any form of prejudice.[14]

Invariants

Suppose that you are about to carry out an extremely important experiment, but you really need a cup of espresso to keep alert. Is it possible that the laws of physics change with time so that the result of the experiment will be different depending on whether you take a break to drink the espresso or to forego it and get on with the experiment? No, because the laws of physics are *time-invariant*, and will not change while you drink your espresso. It is possible that during this time conditions may change—the flux of charged particles from the Sun may increase—and the changed conditions may affect the outcome of your experiment, but the laws of physics will not change. Noether proved that if the laws of physics are in fact time-invariant, than energy must be conserved (recall that mass is a form of energy according to Einstein's famous formula $E = mc^2$). Conservation of energy is a fundamental principle of science that has been extensively verified.

Similarly, suppose that the university asks you to move your laboratory from an airy top floor of a new building to the musty basement of an old building. You might be justified in complaining that the lab is too small, or that the dampness will ruin your experiment, but you would not be justified in claiming that the laws of physics are different in the basement because the laws of physics are *position-invariant*. According to Noether's theorem, this implies that momentum must be conserved; this is a fundamental principle of physics known to every billiard player.

In addition to time- and place-invariants, there are invariants that imply that angular momentum and charge must be conserved.[15]

7 Postmodernist Critiques of Science: Is Science Universal?

Science had always been considered to be value free, in the sense that it describes the natural world *as it is*, not as it *should be* and not as we might wish it to be. Decisions as to what research to pursue and how to apply science were supposed to be left to the conscience of the individual scientist. This changed after World War II because of the revulsion at the perversion of science to further the aims of the Nazis. The value-free approach to science was also called into question by scientists who had worked on nuclear weapons, but who could no longer pretend that they did not share the responsibility for their use with the military and political authorities who actually decided to use them. This led to the *Pugwash Conferences on Science and World Affairs*, whose goal is to involve scientists on issues that affect the security and well-being of the world.[1] Pugwash and its founder Joseph Rotblat (1908–) received the Nobel Prize for Peace in 1995. Similarly, the World Medical Association's Declaration of Helsinki, has guided the ethics of medical research since it was first adopted in 1964.[2]

What all these activities have in common is that they are the efforts of trained scientists to come to grips with the ethical and social aspects of their work. In contrast, science has also been analyzed by sociologists, political scientists and literary critics, who make little or no effort to understand the scientific subjects that they are analyzing. These critiques, known under the umbrella label of *postmodernism*, are seen by scientists as uninformed and pernicious in their effects on the real problems that arise in the relationship between science and society.[3] In this chapter, we shall present some of the issues that arise in this conflict.

The Strong Programme

An extreme form of the claim that social factors influence science, called the *Strong Programme in the Sociology of Knowledge*, was propounded by

David Bloor and others at the University of Edinburgh. Bloor is bothered by the fact that sociological explanations are invoked only when something goes wrong, as in the case of Blondlot's n-rays (see page 70). When cases of scientific fraud are uncovered, like the recent case involving solid-state physicist Jan Hendrik Schön (1970–), sociological explanations are invoked. But according to Bloor, explanations should be "impartial with respect to truth and falsity" and "symmetrical in its style of explanation."[4] Therefore, if French patriotic jealousy caused Blondlot to make a (false) claim that n-rays exist, by impartiality and symmetry, German patriotic jealousy must have caused Wilhelm Röntgen to make the (true) claim that x-rays exist. And if a hypercompetitive environment at the Bell Labs tempted Schön to fraud, then a similar hypercompetitive environment at the IBM Labs must be invoked to explain Binnig and Rohrer's invention of the scanning tunneling microscope (STM).

Scientists, of course, view such claims as nonsense. Röntgen discovered x-rays because they exist, while Blondlot did not discover n-rays because they do not exist. Schön reported that he had built molecular transistors, which he did not, while Binnig and Rohrer invented STMs, which are now routinely manufactured and sold as laboratory instruments. It may be interesting for historians of science to investigate the conduct of Blondlot and Schön, but there is nothing at all impartial or symmetric about the existence or nonexistence of natural phenomena.

Is science universal?

Now we come to the sad part of the story. Science "happened" only once. Every society in the world has developed traditions of art, music, literature (oral and sometimes written), religion, politics, and so on. And every society in the world has observed the natural world and developed generalizations (laws) that were sometimes accurate. But only in one area—Western Europe (primarily Britain, France, German, Italy)—and only in one period—from the sixteenth century on—were the *modern* sciences developed: the science of mathematical physics, the science of the atomic structure of matter, and the science of the biochemistry of living organisms. Furthermore, until relatively recently, almost all scientists were white male Christians.

While many societies (from the Romans to the Zulus to the Incas) have embarked on campaigns of conquest and oppression, the European imperialist expansion was carried out by the same group of cultures and countries, simultaneously with, and aided by, the technology that came with modern science. With the increasing awareness of the injustices perpetrated by European imperialism and of the value of non-European cultures, there is a tendency to lump science as just one more aspect of European culture that was spread by political domination and whose time has come to be pushed off its pedestal in favor of a more democratic approach to describing the world.

In addition, it is undeniable that science has been performed in institutions that practiced stringent discrimination. For many years, membership in the faculty at Oxford and Cambridge Universities was limited to ordained ministers of the Church of England. Over the years, the universities opened their doors first to laymen and Catholics, then to non-Christians, women and nonwhites, but it remains true that most of the pioneers of science do not serve as role models for many people.

We can now ask the question: Does this legacy have anything to do with the *content* of science? In postmodern critiques of science, it does, and science is relegated to the status of just another "discourse" that is amenable to the techniques of literary criticism and sociological analysis. The argument is closely bound up with the classic philosophical issue of *realism* versus *idealism*. A realist is one who believes that the universe actually exists; most realists believe that not only does the universe exist, but also that we can learn (obtain knowledge) about the universe. As we discussed, the problem with realism is that there is no vantage point from which we can check that our knowledge actually corresponds to the universe. Our knowledge is, after all, stored as electrochemical changes in our brain and is obtained from electrochemical impulses transmitted by our senses.

Recognizing the futility of achieving absolute knowledge of the universe, the idealist believes that knowledge is synonymous with our mental constructions, and therefore it makes no sense to talk about the existence of the universe. The universe is simply our knowledge. Taking idealism to the extreme, a *solipsist*, recognizing that our knowledge of any mind outside our own is also transmitted by sensory impressions, concludes that the universe must be a construction of "my" mind alone.

If the real world does not exist, then the claim of science to be describing the real world is nonsense, and science becomes nothing more than another set of precepts similar to a religion. Since democratic societies strive to grant religious and intellectual freedom to all its citizens, any attempt to force people to study science or to support science through taxation or to legislate against what science calls quackery is undemocratic. The privileged position that science has been granted is merely a result of the current distribution of power in a democratic society, and "others" should be "empowered" to live according to their ideas and cultures without being coerced by science.

Nothing, but nothing, drives scientists crazy like these postmodernist claims.[5] They accept that there have been abuses like discrimination and that any remaining vestiges have to be dealt with in all seriousness. But it is axiomatic for a scientist that the universe exists. If it does not, then what is the point? You will never choose to become a scientist if you are not intensely curious about the world. If the planet Jupiter doesn't really exist, why try to theorize about the composition of its atmosphere or spend years conducting measurements with telescopes and spacecraft? The rewards of science (except perhaps for a few like the Nobel Prize) are rather measly compared with those of other occupations. If you are after power, wealth, and fame, it is far better to become a business tycoon, a politician, or a corporate lawyer than to labor away on a professor's salary at a university.

Another reason for not engaging in science if the world doesn't really exist is that science is hard and risky. Suppose that you do have power. If you are an accomplished scientist at a big-name university, you will likely have more access to an accelerator or more funds to buy the latest electron microscope. But if a theory says 100 and your measurements say 1,000, all your years of work and all you power and prestige count for nothing. Many scientists obtain only marginal results after working hard for their entire career, but it is a risk that they take. Nature does not reveal its secrets just because you are high up in the power structure of society, nor will it do so because you deserve it because of past injustices to your group.

The view shared by most, if not by all, scientists is that the universe exists, that it is amenable to scientific research and that science itself is universal, equally accessible to anyone on Earth (and beyond, for that matter) regardless of gender, ethnicity, or culture. It is in the interests of science that the most talented people are provided an opportunity to learn science, to perform scientific research, and to have access to the scientific community, and conversely, opportunity and access should be granted solely on the basis of talent. Impediments to this utopian view should be dealt with for what they are and not as an excuse to propagate idealism and solipsism.

Postmodernism and AIDS

In their attack on postmodernism, Paul Gross (1928–) and Norman Levitt (1943–) discuss in depth (and in a polemical style) the misrepresentations of science perpetrated in the areas of chaos, gender, Afrocentrism, AIDS, the environment, and animal rights. The status of these fields and the attacks upon them do not really form a unified collection: chaos is a real speciality in mathematics that has been overhyped, and there are serious and unresolved scientific issues relating to the environment, whereas the claims that there is a separate and special science for women and Africans border on pseudoscience. We will briefly discuss the postmodern approach to the Acquired Immune Deficiency Syndrome (AIDS) as an example of the pernicious influence of postmodernism in science.[6]

The most important scientific aspect of AIDS is that scientifically there is nothing particularly unusual about AIDS. It is a viral disease, one of many viral diseases, ranging from the almost invariably fatal Ebola to the innocuous "common cold" via diseases that are sometimes deadly and sometimes not, such as influenza, the various forms of hepatitis and the new Severe Acute Respiratory Syndrome (SARS). Nor is it unusual for some people to be more susceptible to a disease than others; Louis Pasteur, a pioneer of microbiology, emphasized 150 years ago the importance of the "terrain" for the growth of a microorganism. Nor is it unusual for some people to respond better to treatment than others, which is why they have those long warning labels in small print attached to every drug. Nor is it unusual for science to find itself unable to cure viral diseases; for the common cold, treatment is limited to paracetamol and hot soup, though for diseases like hepatitis and AIDS, aggressive therapies can significantly retard the progress of the disease. Nor is it unusual for viruses to cause deadly epidemics; in 1918, tens of millions of people died from the influenza pandemic, and it is the memory of that tragedy that drove the aggressive response to the SARS outbreak.[7]

Although AIDS may not be special scientifically, the psychological, social, and political responses to the disease have been unusually strong. Why is this? First, since the disease is almost invariably fatal, it incites fear like no other virus (except perhaps Ebola) can. Second, unlike the 1918 influenza victims who died within a few hours or days, AIDS victims die slowly from serious infections, so that they and their relatives have a long period of time in which to express their anger and despair, clutching at any straws that promise a cure. Third, since AIDS is primarily a sexually transmitted disease (STD), many victims find it embarrassing to seek timely treatment; worse, governments may prefer to avoid dealing with an STD like AIDS for political reasons. As with any STD, many people are reluctant to discuss preventive measures, in spite of the fact that it is much easier to prevent AIDS than it is to prevent influenza or SARS.

Fourth, AIDS was initially identified among homosexual men, amplifying the psychological and social aspects of the disease, with some religious conservatives viewing the disease as divine retribution for sin, while homosexuals viewed the disease from the perspective of their political struggles for equal rights. Finally, the rapid spread of the disease in Africa and Asia, facilitated by political, economic, and social problems (Pasteur's *terrain*), many of which are attributed to European imperialism, led to the rejection of "white" science in favor of traditional practices. Even the valiant efforts of local scientists and physicians to fight AIDS are often hampered by such beliefs.

None of this changes the fact that the science of AIDS is well known and almost universally accepted. AIDS is characterized by the progressive and persistent decline in the number of cells called *CD4+ T-lymphocytes*, which form a vital part of our immune systems. This leaves the victim open to "opportunistic" infections that a normal immune system is usually able to counter. There is a documented case of AIDS from 1959 and the disease may have been in existence for much longer, but the scientific identification of the disease begins with the epidemic of a mysterious immunodeficiency disease in 1981.[8] Relative to most other medical discoveries, the elucidation of the "mystery" of AIDS proceeded quite rapidly, spurred on by the discovery of the *human immunodeficiency virus (HIV)* by Robert Gallo and/or Luc Montagnier in 1984. (Science would not be what it is without a squabble over precedence.)

In chapter 10, we will discuss the concepts of correlation and causation, and the field of epidemiology that studies the prevalence and causation of disease; here it will suffice to say that it is a scientific *fact* that HIV causes AIDS. Recall that in chapter 3 we defined a fact as an observation or theory for which the preponderance of evidence is so overwhelming that no useful purpose would be served by doubting it.

Postmodernists make claims about AIDS that are simply not tenable: AIDS does not exist; poverty causes AIDS; AIDS is merely a new name for well-known diseases; AIDS is an invention of drug companies. Anyone denying that HIV causes AIDS is, in effect, denying the entire content of modern medicine, because this content is based on precisely the same biological and epidemiological methods that show that HIV causes AIDS. The general mechanisms of disease are well known, even if many details are still obscure. Disease is caused by microorganisms, by genetic defects, by toxins, by degeneration of tissues, and by nutational deficiencies—in short, by mechanisms. It is true that people who live in poverty suffer more from disease because of the conditions engendered by poverty. Inadequate nutrition, contaminated water, and crowding make people susceptible to contagion, and illiteracy and poor education leads to sound health advice being unavailable or ignored.

The claim that poverty is the cause of AIDS is similar to the claim that swamps cause malaria, a view that was held for much of history. While it was known that malaria, literally "bad air," occurred near swamps, it could hardly be said that swamps *cause* malaria, because the association is inci- dental in the sense that you can catch malaria where there are no swamps and you can camp near a swamp and not catch malaria. The malaria para- site was first identified by Alphonse Laveran (1845–1922) in 1880, though his work was initially ignored and finally recognized with the Nobel Prize in 1907. The mechanism of malaria—the inoculation by the *Anopheles* mosquitoes of the parasite *Plasmodium falciparum* and its relatives, was finally established in 1897–1898 by Ronald Ross (1857–1932) and Gio- vanni Batista Grassi (1854–1925). Similarly, you can get AIDS without being poor and you can be poor without suffering from AIDS. To treat AIDS merely as a matter of "discourse," "interpretation," and "empower- ment," divorced from consideration of the the underlying viral mechanism is to cause incalculable harm to those who most need the help of modern medical science.

Politics and science

Postmodernists want to politicize science, claiming that science is an activity that is closed to certain groups who need to liberated to engage in their own "ways of knowing." In fact, the history of science shows that science can only thrive in an open society. If rewards are distributed based upon nepotism or party membership, there is no incentive to engage in the intense effort needed carry out scientific research. Furthermore, the self-correcting aspects of the scientific process cannot function; you are hardly likely to perform an experiment that will contradict a theory or experiment published by the son of the duke or the daughter of the party secretary. But even the loftiest professor at Harvard or Cambridge is fair game for a graduate student at the most obscure university.

Great efforts are made to ensure universal access to science; journal subscriptions are sold openly and are available in university libraries that usually have rather liberal browsing policies. Most scientific conferences are open to all who wish to attend. When decisions must be made, for example, which articles to publish, who gets to present a paper at a conference, and which projects are funded, multiple, anonymous referees are consulted. The ability to assert hegemony over scientific results is quite limited and short-term.

When the open exchange of ideas breaks down, the results for science can be traumatic, as can be seen in the Lysenkoism affair. During all the years that the Soviet Union existed, their physical science and mathematics were first-rate. Scientists were granted special privileges, both material and in terms of access to foreign journals and travel, in return for nominal support of the Soviet ideology. In the biological sciences, however, the Communist Party decided to meddle in the *content* of a scientific theory. Trofim Denisovich Lysenko (1898–1976) propounded a version of Lamarck's theory of evolution, claiming that characteristics acquired during the lifetime of a living organism can be passed on to its offspring. He held this theory despite all the evidence that supported Darwin's theory of evolution in its synthesis with Mendelian genetics; according to these theories, variation is unpredictable and variants are selected for based upon reproduction advantage in an environment.

Lysenkoism was useful to the Soviet authorities because it implied that political control that changed the behavior of a person could in time produce offspring who would demonstrate the desired behavior from birth. Soviet society was closed and dictatorial, and dissenters from Lysenkoism were silenced, usually by execution or imprisonment in Siberia. Since you can't rebuild a scientific community in a day, biological science in Russia took decades to recover from this political meddling.

A more gruesome episode in which political ideology was mixed with science was the support given to Nazi ideology by German psychiatrists and anthropologists. Not only did they provide the theoretical background for the well-known genocide of Jews, but they also helped carry out forced sterilization and later mass murder of mentally ill patients and others who were entrusted to their care. An account of this unholy alliance is truly spine-chilling and conclusively shows that politics and science should not be mixed.[9]

Both of these travesties of science were carried out in closed societies, in which the "sciences" in question were not exposed to the checks and balances inherent in the international exchange of ideas. Soviet biologists ignored the progress in genetics that had been made in other countries, and papers describing Nazi experiments would never have been accepted for presentation at international conferences. It is therefore highly paradoxical that postmodern criticism insists on fragmenting science into feminist science, ethnic science, and so on, because it is precisely in the open atmosphere of international science that values such as respect for other cultures can be nurtured.

Japan: A case study in the universality of science

To counter the postmodernist demand for different sciences for different groups, it is worth recounting two historical episodes, one of a society and one of a person, that show that science is universal, even though scientific processes can be affected by the characteristics of a society.

If there ever was a society that is a candidate for having a different "way of knowing" and a different science, it is certainly Japan before the mid-nineteenth century. A long period of isolation had left Japan with a society whose concepts of culture and politics were totally different from those of European society. Furthermore, although Japanese society had been more technologically advanced than feudal European society, it had stagnated in the seventeenth century just as modern science began to be developed in Europe. The Japanese were astounded by the power of the technology displayed by Commodore Matthew Perry (1794–1858), who showed up on their doorstep in 1853 in a fleet of American warships. If they had been good postmodernists, they would have insisted on retaining their own ways of knowing and totally rejected European culture and science.

Instead, beginning with the political transformation called the Meiji Restoration in 1867, Hirobumi Ito (1841–1909) and others decided to drive Japan into accepting Western "ways of knowing" the natural world, without significantly modifying other aspects of their culture. Japanese students were sent to Europe and America to study science and technology, and within a mere forty years, in 1905, the Japanese destroyed the Russian fleet at Tsushima and defeated the Russian army at Port Arthur, using the Western science and technology they had learned. Unfortunately, Japanese culture included deeply embedded elements of ultranationalism and militarism, so that from the 1930s they used the European technology that they had assimilated to make war first on China and then on the rest of the world. Since World War II, however, Japan has renounced militarism and has become a leader in science and technology, all the while retaining, as far as possible, their own culture and social institutions. It is hard to explain the rapid scientific and technological achievements of the Japanese, unless one accepts the realism of science, both in the sense that the universe exists and in the sense that scientific knowledge is universally accessible, and that its content does not depend on culture.

Ramanujan: A case study of the universality of mathematics

Srinivasa Ramanujan (1887–1920) grew up in a poor family near Madras in southern India. As a young student, he found a book of raw mathematical formulas like the tables of integrals and trigonometric formulas that you find in the appendices of math textbooks. Ramanujan first proceeded to recreate all the mathematical theory behind these formulas, and then extended them and invented new mathematical theories. He was eventually brought to Cambridge University by the famous mathematician Godfrey Harold Hardy (1877–1947), although initially Ramanujan was reluctant to travel because it was inconsistent with his Hindu-Brahmin beliefs. During his sojourn in England, Ramanujan maintained a strict vegetarian diet, but soon after his arrival, World War I broke out, making it impossible to import familiar foods from India. Weakened by the cold climate and dietary difficulties, Ramanujan fell ill; he returned to India after the war and died at the age of 32.

Ramanujan's short career is an ideal case study in which to check if mathematics is universal or not, and if there really are different "ways of knowing." In terms of culture, what could possibly be common to a white, Christian, English-speaking, meat-eating European and a black, Hindu-Brahmin, Tamil-speaking, vegetarian Asian? Yet, though Hardy found some of Ramanujan's mathematics strange, it was consistent with "European" mathematics, and Hardy, one of the greatest European mathematicians, judged it as incredibly original and brilliant. The strangeness was attributed to Ramanujan's lack of formal education in European mathematics; as Hardy said: "Ramanujan had none of Landau's weapons at his command; he had never seen a French or German book; his knowledge even of English was insufficient to qualify for a degree." But there was no question of his brilliance: "It is sufficiently marvelous that he should have even dreamt of problems such as these, problems which it had taken the finest mathematicians in Europe a hundred years to solve, and of which the solution is incomplete to the present day."[10]

Ramanujan's mathematics was understood, even if it took some effort to unravel his unusual presentation, and to this day, mathematicians are working on problems and theories that Ramanujan initiated. Most mathematicians view mathematics as universal; to quote Hardy again: "My belief is that all mathematicians think, at the bottom, in the same kind of way,...."[11]

Whatever the faults of science, one of its greatest virtues is its universality, and hence its accessibility to anyone regardless of personal background. That is why the postmodernist criticism of science, unwittingly ignited by Thomas Kuhn, is seen by scientists as extremely unfair and damaging *precisely to those* who have the most to lose because they come from social groups whose participation in science has been marginal. Postmodernism, which champions the right of marginalized groups like women and nonwhites to different "ways of thinking," is paradoxically guiding future Emmy Noethers and Srinivasa Ramanujans into pointless activities that will marginalize them even further, instead of encouraging them to strive for achievement in activities where their origin is truly irrelevant.

Sokal's hoax

Let us conclude this chapter by presenting the affair of *Sokal's hoax*, a recent incident that engendered vociferous debate on science and postmodernism. In 1996, Alan Sokal (1955–), a physicist at New York University, published an article called *Transgressing the Boundaries: Toward a Transformative Hermeneutics of Quantum Gravity* in the journal *Social Text* devoted to postmodernist cultural studies.[12] Simultaneously, Sokal published another article in which he confessed that the article in *Social Text* was a parody. Sokal asked himself:

> Would a leading North American journal of cultural studies—whose editorial collective includes such luminaries as Fredric Jameson and Andrew Ross—publish an article liberally salted with nonsense if (a) it sounded good and (b) it flattered the editors' ideological preconceptions?[13]

From a postmodernist perspective, the article certainly sounds good and ideologically correct. Here is one (!) typical sentence:

> It has thus become increasingly apparent that physical "reality," no less than social "reality," is at bottom a social and linguistic construct; that scientific "knowledge," far from being objective, reflects and encodes the dominant ideologies and power relations of the culture that produced it; that the truth claims of science are inherently theory-laden and self-referential; and consequently, that the discourse of the scientific community, for all its undeniable value, cannot assert a privileged epistemological status with respect to counter-hegemonic narratives emanating from dissident or marginalized communities.[14]

The article is certainly liberally salted with nonsense. Not being a physicist, I can't vouch for Sokal's claims about his central topic of quantum gravity, but I was quite amused by the following claim:

> It is not clear to me that complex number theory, which is a new and still quite speculative branch of mathematical physics, ought to be accorded the same epistemological status as the three firmly established sciences cited by Markley.[15]

All college students of mathematics, science, and engineering, and even many high-school students, know that complex number theory is not speculative and is routinely used in a vast array of applications. Hardly new, it is a venerable field of mathematics whose basic theory was worked out in the early nineteenth century by Carl Friedrich Gauss (1777–1855) and Augustin Louis Cauchy (1789–1857). As Sokal writes:

> Evidently the editors of *Social Text* felt comfortable publishing an article on quantum mechanics without bothering to consult anyone knowledgeable in the subject.[16]

In an editorial response, Bruce Robbins and Andrew Ross, the editors of *Social Text*, wrote that they decided to publish the article in the interests of diversity, although they claim that if subjected to academic review in cultural studies it would not have been publishable:

> As the work of a natural scientist it was unusual, and, we thought, plausibly symptomatic of how someone like Sokal might approach the field of postmodern epistemology, i.e. awkwardly but assertively trying to capture the "feel" of the professional language of this field, while relying upon an armada of footnotes to ease his sense of vulnerability.[17]

They certainly did not feel that a scientific review was required, for the simple reason that they are not doing science:

> Sociologists of science aren't trying to do science; they are trying to come up with a rich and powerful explanation of what it means to do it. Their question is, "What are the conditions that make scientific accomplishments possible?" and answers to that question are not intended to be either substitutes for scientific work or arguments against it.[18]

A further complaint by postmodernists is that scientists are attacking caricatures of sociologists of science and cultural critics by accusing them of rejecting the existence of the real world:

> Like Gross and Levitt, he appears to have absorbed these critiques only at the level of caricatures, and has been reissuing these caricatures in the form of otherworldly fanatics who

deny the existence of facts, objective realities, and gravitational forces. We are sure Sokal knows that no such person exists, and have wondered why on Earth he would promote this fiction.[19]

The problem with these protestations of innocence is that postmodernism *does* seem to treat science in a superficial and relativist manner. The charge of relativism is hard to refute when the editors of *Social Text* agree to publish an article which seeks to accord quantum mechanics the same epistemological status as "counter-hegemonic narratives emanating from dissident or marginalized communities." A sociologist can certainly study the behavior of physicists without a professional knowledge of quantum mechanics, but it is hard to conceive of how that could be done without a reasonable level of understanding of the subject. It is as if one undertook to conduct literary criticism of Shakespeare's historical dramas without a reasonable familiarity with English history in the fourteenth and fifteenth centuries, or if one undertook to study the sociology of a symphony orchestra without ever having listened to the music of Beethoven.

The Sokal affair illuminated the problems of the current relationship between science and the humanities. To the extent that researchers study the social, political, and literary aspects of the practice of science, their efforts result in interesting and important findings, and most scientists are willing cooperate or at least tolerate their activities. But if nonscientists make claims about the *content* of science, and if these claims are based upon only the most superficial acquaintance with that content and with the ways in which scientific knowledge is created and validated, then scientists will certainly respond with indignation and disgust.

* * *

A scientific theory is concerned with natural phenomena. If "nature" doesn't exist, or if it is merely a construction of the human mind, then there is really no point to doing science, except perhaps to the extent that it brings us new gadgets to make our lives easier. Certainly, if science is just another "discourse," there is no intellectual justification for spending long hours laboring over the mathematical details of a theory or painstakingly carrying out a delicate experiment. The scientific point of view is that such claims of postmodernism are a travesty that comes from mixing ideology and politics with science.

JUDAH FOLKMAN: PERSISTENT OUTSIDER

Judah Folkman (1933–) is a modern-day Alfred Wegener (see page 198), a scientist from an allied discipline proposing a radically new theory that was initially totally rejected and only later accepted into the mainstream of the research in the field. If I seem to be emphasizing these figures at the expense of others whose accomplishments are within the fields they trained in, it is for a reason. So often pseudoscientists point to these figures as evidence of the persecution visited upon every new theory. The implication is that just as Wegener and Folkman were eventually vindicated, so will come the day when scientists will apologize to astrologists and homeopaths. They totally ignore the delicate difference—Wegener and Folkman were dedicated scientists who realized that their theories would only be accepted if they were backed up by evidence, and they spent decades amassing the evidence using methods that were convincing to other scientists.

Judah Folkman trained as a surgeon at Children's Hospital of the Harvard Medical School. In the early 1960s, while working at a research lab of the US Navy on the preservation of hemoglobin, he stumbled across a strange phenomenon: cancerous tumors would stop growing if they were not copiously supplied with blood. He proposed that tumors secreted a substance that encouraged *angiogenesis*, the growth of new blood vessels. The implication was that a medication to inhibit this substance could be used to strangle tumors.

Folkman was initially laughed at, or worse. His proposal was totally alien to the prevailing fields of cancer research, which were cellular and molecular biology. In retrospect, Folkman believes that it was precisely because he came to cancer research from the outside that he was able to see the importance of his discovery. His extensive experience as a surgeon excising bloody tumors gave him a perspective that was lacking in scientists who worked only with preserved pathology specimens. Like Henri Becquerel understanding the importance of a serendipitous fogged photographic plate, Folkman the physician understood the importance of a tumor starved for blood.

It took more than a decade, but eventually scientists at Folkman's lab were able to isolate proteins that encouraged angiogenesis. It took another decade to find the first inhibitors, some of which are now in clinical trials on cancer patients. The evidence was sufficient to convince most of his critics, and research on angiogenesis is no longer outside the mainstream of medical research.[20]

Judah Folkman's work also demonstrates that scientific research has a future, even in the absence of Kuhnian revolutions. There was nothing revolutionary about the new theory in the sense that it involved well-known principles of biochemistry, but by any definition, the new theory was surprising, interesting, and useful.

8   Science and Religion: Scientists Just Do Science

Is there any connection between science and religion? In this chapter, we will explore possible answers to this question. Since religion predates science by thousands of years, and since scientific claims ostensibly contradict some religious claims (or at least certain interpretations of these claims), the relationship between the two has been stormy and has fascinated people since the beginning of modern science. Entire books have been written in support of every possible viewpoint, so we can only hope to briefly survey this debate.

So, is there any connection between science and religion? There are roughly two answers to this question: yes and no. Many, if not most people, both religious and nonreligious, agree that there is a connection. This is particularly true for people who hold the more extreme positions on either side of the coin. We will start by analyzing these views, and then consider the alternative, namely, that there is little or no connection between science and religion, a viewpoint that is quite prevalent among scientists, both religious and nonreligious.

Religious claims about natural phenomena

For most religious people, religion is not a menu that you pick and choose from, but rather an all-encompassing framework within which life is lived. The source of knowledge about the framework is usually a holy scripture, interpreted by members of the clergy of the religion. Clearly, if you are religious in this sense, it may be difficult to accept that there exists an alternate source of true knowledge like science, which claims that reliable knowledge of the universe can be obtained by empirical and rational investigation.

This will hold to the extent that the religion makes claims about phenomena that science sees as natural phenomena amenable to scientific investigation, and to the extent that the scientific claims contradict the religious claims. To take a fanciful example, if the entire content of a religion were limited to a description of life on planets in the Andromeda galaxy, then the potential for conflict between that religion and science would be minimal. Conversely, since science (currently!) has nothing whatsoever to say about life forms in the Andromeda galaxy, scientists could equally well accept or reject the religion.

It is not surprising, however, that religions make claims concerning our lives as humans. If, as religious people believe, we were created for some divine purpose, or if, as nonreligious people hold, religion was constructed by humans to supply psychological, social, and political needs, in either case, it is obvious that religion will make claims that are relevant to our lives.

The existence of a conflict between science and religion depends on the existence of contradictory claims. For example, I am not aware of a serious conflict between chemistry and religion, because neither makes claims that are likely to contradict claims from the other field. With a few possible exceptions (like the transformation of water into wine, the miracle attributed to Jesus), religious texts do not contain significant material on, say, the reaction of acids and bases. Similarly, there should be little conflict between religion and fields such as physical optics or superconductivity.

Conflict between science and religion exists because there are fields in which there is significant overlap between the claims of science and those of religion. There are now well-established scientific theories describing the origin and subsequent development of the universe, the origin and subsequent development of the planet Earth, and, most significantly, the development of life, including human life. Most religions contain a description of the creation of the universe and the Earth, as well as of the origin of humans, so conflict is unavoidable. Potential conflict between science and religion exists, above all, in some areas of physics, and in geology and biology, because these sciences are ineluctably bound up with theories that provide natural, nonreligious explanations of the origins and development of the world as we experience it.

Before the rise of modern science, religion had taken upon itself to explain phenomena that could not otherwise be explained. If you were sick, it was because God was punishing you for your sins; if there was a drought, it was almost certainly because God was punishing the community for tolerating a sinner in its midst. However, as modern science has offered explanations for more and more natural phenomena, the need for religious explanations has diminished. Sickness caused by microorganisms and drought caused by a change in an ocean current have nothing to do with our sins.

Furthermore, if religion is an aspect of human culture, naturalistic explanations can be offered for its existence. A political explanation is that religion was invented by political leaders to solidify their control over the populace. If a king rules "by the Grace of God," then his commands must be obeyed on pain of severe punishment, both in this world and in the next.

An alternative naturalistic explanation of religion comes from evolution. With the evolution of consciousness, significant selective advantage could accrue to individuals with minds adapted to religious thoughts. Such an individual might be better able to bear the pain of a hunting injury and bring food to his children if he believed that God is protecting him. The evolution of human behavior is a new and controversial field, but it is clear that to a nonreligious person there is no a priori reason why the origin of religion should be excluded from scientific study.

To summarize, there is a potential conflict between science and religion to the extent that religion makes claims about phenomena in the world that are studied by science and conversely. Many people, both religious and nonreligious, hold that this situation does in fact exist. Since we cannot step outside the universe in order to verify what the truth really is, this conflict can never be conclusively resolved.

Nonoverlapping magisteria

People who look for a way to reduce the friction between science and religion claim that there is no inherent conflict between the two. Many such people are bona fide scientists, some of whom are religious people looking to reconcile their religious beliefs with their scientific profession, and others are secular scientists who seek a modus vivendi with religion, especially in countries where the majority of people identify themselves as religious. Rapprochement is also sought by members of the clergy who see no advantage to a confrontation of religion with the overwhelming evidence for the success of science in describing the modern world.

One view that advocates rapprochement between science and religion has been called *nonoverlapping magisteria (NOMA)* by the late paleontologist and science writer Stephen Jay Gould (1941–2002).[1] Science and religion are deemed to have authority in different areas of human life—science describes the natural world, while religion prescribes how humans should live their lives in terms of ethics and morals. Galileo, who believed that the book of nature and the Bible were two separate books, is supposed to have said that "the Bible tells us how to go to heaven, not how the heavens go."

Gould was forthright that his proposal derived from the political and moral views that he held:

> I believe, with all my heart, in a respectful, even loving concordat between our magisteria—the NOMA solution. NOMA represents a principled position on moral and intellectual grounds, not a mere diplomatic stance.[2]

Even such a conservative religious figure as Pope John Paul II has written that the Roman Catholic Church has no objection to the study of evolution as an explanation for the origins of the human body, as long as members of the Church believe that at some point God infuses the body with an immortal soul.

Yeshayahu Leibowitz (1903–1994), a scientist and philosopher who practiced Orthodox Judaism, expressed a similar sentiment in extremely forceful language:

> There is no relationship of belonging or mutual dependency between scientific knowledge and decisions about values; there is no contact and there is no conflict. . . . What can the immense achievement that is science—which is accepted by everyone who understands it—contribute to these decisions on values? Science cannot contribute anything at all, because concerning the problem addressed by these decisions—to be a decent person or a rascal, a patriot or cosmopolitan, a believer or an apostate—not only does science have nothing to contribute, but these questions cannot even be posed, because these concepts simply do not appear in the lexicon of science.[3]

NOMA is a successful philosophical position in that it enables a scientist to live comfortably in both worlds. During the week, he can perform the most advanced research in molecular biology or astrophysics, and then on Friday or Saturday or Sunday, he can perform his duties as a member or even as a leader of a religious group.

The NOMA position does seem to demand a level of intellectual sophistication that is beyond the reach of ordinary people who want simple answers to questions like: "How did we get here?" and "Why are we here?" Furthermore, most religious people need a *personal* God who will reward them for doing good and punish them for doing bad, a God who will answer their prayers. Similarly, many nonreligious people reject NOMA because they regard it as an attempt to evade dealing with religion in scientific terms. It is as if science investigates *all* phenomena in the universe, except that it is forbidden to investigate religion, which as a human activity is no different from other activities that are the subject of scientific investigation.

Reconciling science and religion

Another approach to reconciling science and religion tries to find some interpretation in which the results of science and the tenets of religion are consistent. This is particularly true in the United States where the Constitution is interpreted as prohibiting the teaching of religion in public schools. Therefore, if a religious tenet could be shown to be scientific, it would be possible to introduce it into a school curriculum. *Scientific creationists* attempt to interpret scientific observations and theories as supporting a literal reading of the biblical story of creation, but this requires the rejection of so much well-established science that it is not seriously considered by scientists.

Scientific creationism is the outgrowth of a particular political situation, but it is an example of approaches that attempt to find room for a divine being within the scientific world. These approaches, often called *natural theology*, seize upon the obvious fact that since science doesn't explain everything, some things can be explained by religion. Obvious examples of unexplained phenomena are the origin of the universe and the nature of human consciousness. The big bang theory is widely accepted as explaining the mechanism of the origin of the universe, but it does not answer questions like: "Why is there something instead of nothing?" and "What caused the big bang?" Science has no way of answering such questions, not currently, and probably not in the foreseeable future. The religious answer that God created the universe is thus fully consistent with the big bang theory.

Similarly, since science cannot (currently) explain consciousness nor (currently) prove if free will exists or not, there is no inconsistency in believing that humans possess an immortal soul instilled by God at birth and destined to be judged after death. It is even possible to integrate natural theology with a personal god. One could hardly have a religion based upon a holy book that began: "God created [place here the equations of quantum mechanics and relativity]," followed a few chapters later by: "God created [place here the three billion sequence of nucleotides of the human genome]" (sidebar). For one who believes in some form of natural theology, the holy scriptures and religious rituals can be interpreted as metaphors whose scientific truth is irrelevant.

Natural Theology

And God said

$$\epsilon_0 \oint \mathbf{E} \cdot d\mathbf{S} = q \qquad\qquad \oint \mathbf{B} \cdot d\mathbf{S} = 0$$

$$\oint \mathbf{B} \cdot d\mathbf{l} = \mu_0 \left(\epsilon_0 \frac{d\Phi_E}{dt} + i \right) \qquad \oint \mathbf{E} \cdot d\mathbf{l} = -\frac{d\Phi_B}{dt}$$

and there was light.

Natural theology would claim that God created the universe and then the creation and extinction of millions of species of life on Earth was carried out by the natural process of evolution. For a nonreligious person, it is incredible that God would have engaged in such complex and unnecessary activities if the entire purpose were only to create humans to serve him. Furthermore, natural theology is problematical because it is a moving target: as long as there is a gap in scientific knowledge, God can be invoked to close the gap, but as soon as scientific knowledge has advanced and closed the gap by providing a natural explanation, the "god hypothesis" has to be further constrained. This is why natural theology of this form has won the derisive label of "God of the Big Bang" or "God of the Gaps."

Science is incomplete, and the uncertainty is greatest at those points of the most intense philosophical curiosity for us like the meaning of life. Furthermore, our existence in this world seems insignificant within the vast extent of space of and time. Therefore, nonreligious people have to come to terms with living in a world full of uncertainty and unknowns. Nevertheless, many people prefer facing the uncertainty, rather than believing in a certainty makes no sense to them.[4]

The slippery slope

Perhaps you are wondering why the biographical vignette following this chapter portrays a computer scientist. What possible connection could there be between computer science and the issue of science and religion? The superficial answer is that there is no connection, which is one reason why many religious people are attracted to computer science. Computer technology is a boon to the dissemination of information and to the creation of virtual communities, and religions have eagerly accepted the use of this technology. Furthermore, you can ostensibly participate in the use of computers without being troubled by the controversies that are engendered by subjects like evolutionary biology, geology, and cosmology.

Nevertheless, should creationists win their battle with evolution, there is no reason why they would stop there, and the slippery slope can even bring us to computer science. There are many beautiful theorems in computer science that show that there *do not exist* algorithms (or efficient algorithms) for solving certain problems. The most famous result was proved by Alan Turing (1912–1954), who showed that it is impossible to write a computer program that decides if the execution of any arbitrary program terminates or not. Perhaps in the future, we shall have to add stickers to our textbooks: Turing's theorem does mean that God cannot write such a program (sidebar page 138).

* * *

A scientific theory attempts to explain natural phenomena. If there is something supernatural, beyond nature, by definition it is not a concern of science. If a religion concerns itself with supernatural phenomena, by definition it will not conflict with science. However, many, if not most, religions do make claims about natural phenomena, leading to conflict if its explanations and predictions are not consistent with the scientific ones. People, including scientists, hold a wide range of religious and philosophical beliefs. However, all reputable scientists accept the main body of scientific theories, observations, and experiments, and are able to work together regardless of their personal religious beliefs or lack thereof.

A Message From the Alabama State Board of Education[5]

[to be pasted in all physics textbooks]

This textbook discusses gravitation, a controversial theory some scientists present as a scientific explanation for the motion of bodies such as projectiles and the solar system. No one is present within the Earth or on the Sun or the planets. Therefore, any statement about the forces acting upon them should be considered as theory, not fact. The word "gravitation" may refer to many types of motion. Gravitation describes the motion of everyday objects. (Footballs, for example, may "gravitate" towards the surface of the Earth.) This process is microgravitation, which can be observed and described as fact. Gravitation may also refer to the effect of one heavenly body on another, such as Pluto on Neptune. This process, called macrogravitation, has never been observed and should be considered a theory. Gravitation also refers to the unproven belief that certain mathematical equations describe the structure of space and time. There are many unanswered questions about the theoretical force called gravitation, which are not mentioned in your textbooks, including:

- What does it mean for an object to have "mass"?

- Why is the theory of gravitation so hard to reconcile with quantum mechanics?

- If the gravitational force is transmitted by waves or by particles, why have they never been observed?

- How did you and all objects in the universe come to possess the ability to attract other objects with the same gravitational constant?

Study hard and keep an open mind. Someday you may contribute to the theories of how objects move.

EDSGER DIJKSTRA: BRINGING MATHEMATICS TO COMPUTING

The term "hacker" conjurers up an image that epitomizes a computer programmer: a person—usually a young male with limited social skills—doing battle with the computer through long dark nights until finally the last bug is slain, though rarely is a princess waiting for the hero. But there is another culture that views programming as a mathematical activity, based upon abstractions, precise definitions, and rigorous proofs. The most eloquent crusader for a mathematical approach to programming was the late Edsger Dijkstra (1930–2002).

Contrary to the bad-schoolboy myth attributed to many pioneering scientists, Dijkstra received excellent grades and went on the study theoretical physics. When he applied for a marriage license, Dijkstra tried to list his profession as "computer programmer," but this occupation was not yet recognized by the municipal authorities in Amsterdam! Fortunately, "theoretical physicist" was. Like many pioneers in computer science, he started his career performing scientific calculations on a computer, only to quickly focus his attention on the limitations of the programming process itself and on ways to put it on a firm theoretical base. Most programmers see programming as an ending sequence of puzzles to be solved, but Dijkstra was one of those who rose above solving puzzles in order to create mathematical models of programming.

Dijkstra's short letter entitled "Go To Statement Considered Harmful" inaugurated an effort to clean up programming languages and programming techniques from their unstructured anarchy.[6] More important was Dijkstra's forceful demonstration of the importance of mathematically verifying the correctness of algorithms and programs. We have become so used to bugs in programs that we hardly dare imagine a world in which programs work the first time they are turned on. Dijkstra did. His techniques for crafting programs together with their proofs of correctness spawned a whole field of research, though there are many who regard this as irrelevant to "real" programming. As more and more critical systems are controlled by computers, these techniques will be applied, though they will perhaps be forever beyond the grasp of a hacker.

Dijkstra's innovative ideas granted him the stature and freedom to fulfil the role of gadfly to the computer science community. An equally long stream of over 1,300 letters and technical notes flowed from his pen (literally, he never used a word processor!), and contained not just technical ideas, but also continued advocacy of the place of mathematics in computer science.[7]

In the 1980s Dijkstra moved from the Netherlands to the University of Texas, where he attempted to preach his gospel of the mathematical basis of programming to American computer scientists.

I used to say that computer science was such a young field that its pioneers are still alive. Sadly that is no longer true: Edsger Dijkstra passed away on August 6, 2002.

9   Reductionism: The Whole Is the Sum of Its Parts

A mountain climber in the Alps reaches a high plateau, where he finds a shepherd munching on a large hunk of cheese while guarding his flock. Tired and hungry, the climber asks if he could buy a piece of cheese, but the shepherd tells him regretfully that he has none to sell. Undeterred, the climber proposes a challenge: if he can guess the exact number of animals in the flock, will the shepherd reward him with a piece of cheese? The shepherd agrees, and after glancing at the flock for a few moments, the climber reports, "There are exactly 329 sheep in the flock."

Astonished at the precise answer, the shepherd breaks off a generous hunk of cheese and gives it to the climber. As the climber gratefully gnaws at the cheese, the shepherd looks at him and says, "I'll bet 100 Swiss francs that I can guess what your profession is."

"You're on," replies the climber.

"You must be a molecular biologist," says the shepherd.

"Why, yes I am, but how did you figure that out?" asks the amazed climber.

"Simple," replies the shepherd, "those animals are goats, not sheep!"

The author of this story is accusing the molecular biologist of the grave cultural "sin" of being a *reductionist.* A reductionist is supposed to be a scientist who studies the universe by disassembling everything into its pieces, thereby losing sight of the holistic whole. Molecular biologists are assumed to be guilty of reductionism because they look upon life as composed of dry strings of the letters A, C, G, T (representing the nucleotides), instead of appreciating the holistic nature of life: the beauty of a flock of goats grazing peacefully in a lush Alpine meadow, lulled by the gurgling brook and caressed by a sensuous breeze.

Reductionism and mechanism

Science attempts to explain a natural phenomenon in terms of a mechanism. The nature of an explanation or a mechanism is such that it necessarily involves lower-level entities and concepts than the phenomenon under discussion. We say that the phenomenon is *reduced* to lower-level entities and concepts. This is so trivial that it would not be worth belaboring, except for the misunderstanding it engenders concerning the process of science. For example, if you ask me to explain why my spaghetti sauce tastes so good, I would offer an explanation by *reducing* the question of taste to a detailed list of ingredients; for example, I would tell you that I use lots of garlic and thyme. Any time you ask for a recipe, you are engaged in reductionism, that is, you have abandoned the holistic experience of tangy taste, pungent smell, and velvety texture for the dry list of cloves of garlic and sprigs of thyme.

Similarly, a scientist searching for explanations is necessarily going to reduce a phenomenon to lower-level concepts and entities. Of course these lower-level entities will themselves become objects of curiosity and demand explanation in terms of a still lower level. If you are curious to know why sheep chew cud, you have no choice but to reduce the problem from a study of sheep as individual animals to a study of the anatomy and physiology of the internal organs of sheep. In turn, the anatomy and physiology can be explained by reducing them to the muscular structure of the complex ruminant stomach and the biochemistry of the digestive secretions. Reductionism runs rampant and stops only at the level of the elementary particles of physics.

Abstraction

The reverse of reduction is *abstraction*. The essence of abstraction is "forgetting" detail, that is, ignoring lower-level details in order to concentrate on properties that are of interest at a higher level. These properties would be obscured by overwhelming detail if the description were to remain at the lower level. Take music as an example. A composer such as Ludwig van Beethoven writes symphonies in an abstract notation that has become conventional in the world of music. The notation specifies the pitch and relative duration of the notes, and includes rough indications of tempo and volume. Even so, the musical scores for the symphonies form thick volumes. Composing would be totally impractical if there were no abstraction, for then instead of simply specifying by a single symbol that the oboe is to play a high C, the composer would have had to specify exactly which holes on the oboe must be covered and which keys must be pressed. The oboe player is as dependent as the composer upon the abstraction; imagine how impractical it would be to play the instrument if every note in the score were denoted by a diagram showing all the holes and keys.

Abstractions, of course, must be implemented if they are to be useful. An artisan facilitates the implementation of the musical abstraction by manufacturing an oboe from wood and metal, using, of course, knowledge at a still lower level, for example, knowledge concerning the relation between the length of a vibrating column of air and the pitch of the tone it emits. If we did not have such abstractions, there would be no way of manufacturing a musical instrument that would be appropriate for all compositions; instead, we would need a holistic approach to music where the design of the instrument takes into account the composition for which it is to be used.

Explanation requires reduction to a lower level, while abstraction is essential in order to function at a higher level. But these qualifiers "lower" and "higher" are just technical adjectives and have no value judgment associated with them. The composer, the conductor, the musician playing the oboe, and the artisan who built the oboe each work at different levels, but they are all essential to the production of a moving musical experience. Talented artisans like Antonio Stradivari who craft musical instruments are held in awe no less than talented composers like Beethoven, even though they work at a "lower" level in the chain of reductions.

Abstraction and reduction in science

If you aren't willing to engage in abstraction and reduction, you are not going to be a scientist. If you aren't willing to work with abstractions, you will be overwhelmed by irrelevant detail, and if you aren't willing to work with reductions, then you won't be able to produce any interesting scientific theories. Scientists must work at a boundary between levels, studying abstractions and then explaining and predicting phenomena in terms of a reduction. A zoologist must be intimately familiar with the anatomy and physiology of animals; a physiologist with biochemistry; a biochemist with organic chemistry; an organic chemist with quantum mechanics. A good scientist is one who can switch levels with ease, though always being cognizant of the different levels at which she is working. The joke about the climber and the goats is not entirely farfetched, in that a scientist can become totally wrapped up in her research and forget about the outside world. But this is a personal trait or an outcome of the intense concentration needed to perform scientific research, not an essential aspect of science itself.

The praise for holism as opposed to the pejorative connotation of reductionism is almost certainly the result of the behavior of certain physicians. Those in highly technical fields like surgery can become so immersed in the technical details that they "see" the disease and forget to look at the patient. If the patient feels neglected he is likely to search for caregivers who specialize in listening sympathetically to the patient. But that doesn't change the fact that the practice of surgery demands extensive knowledge at a very specific level of abstraction, namely, the anatomy and physiology of organs and tissues, as well as mastery of techniques and technology. A surgeon's talents at a higher level of abstraction (his rapport with patients) and at lower levels of reduction (the biology of the cell) are of less importance.

Limitations upon the ability of any single individual lead to specialization. You pick your specialty, study some of the relevant abstractions and reductions, but mostly remain at the level you have chosen. There is no value judgment inherent in any particular level.

Idealization by abstraction and reduction

There is a story about a physicist who was given a grant to investigate the aerodynamic properties of the saddle. The grant was funded by a rich breeder of racehorses, who was convinced of the ability of science to help him win more races. After getting his grant extended for a second and then a third year, the physicist triumphantly announced to his sponsor: "I've solved the problem—for a one-dimensional horse!"

In the process of searching for explanations, it is rarely, if ever, possible to explain every detail of a complex system. The reduction of a problem will start out by identifying small, often idealized, parts, such as point masses, individual molecules, or single cells. Even when an explanation is found, it is not necessarily applicable to an entire system like a star or a horse. Instead, abstraction must be performed, and in the process of abstraction, new tools are invariably needed.

Consider the status of meteorology as a science. Weather is a macroscopic manifestation of the motion of the innumerable molecules that comprise our atmosphere. The underlying scientific theories are well established: statistical mechanics and the thermodynamics of the absorption of radiation from the Sun and the heat exchange between the atmosphere and the oceans and landmasses. An extreme reductionist might deny an independent scientific status to meteorology, claiming that *in theory* meteorology can be reduced to physics; *in practice*, of course, the number of molecules in the atmosphere is so huge that it is completely impossible to perform the reduction. The use of the term "in practice" is a bit misleading since it seems to imply that with slightly more powerful computers, we could reduce meteorology to physics. However, it is simply impossible to measure the properties of each individual molecule, and the computational power needed is so immense, that "in practice" turns out to be "in theory" as well.

Meteorologists know much about the mechanisms that are responsible for the weather: the chemical and physical properties of molecules of oxygen, water, nitrogen, and so on. Without the knowledge of these mechanisms, weather would be just as mysterious to us as it was when people believed that a rain dance would cause rain. However, meteorology exists as a distinct science because it is not possible to study the weather in terms of the individual molecules. More abstract concepts and techniques are needed, for example, a *front* as an abstraction for a pressure and temperature gradient in the atmosphere. These concepts can be reduced to statistical mechanics (which in turn is based on the properties of individual molecules), but meteorology is not a branch of statistical mechanics. Meteorology is not more "holistic" nor less "reductionist" than statistical mechanics; it is simply a different science, working at a different level of abstraction.

It is often claimed that as science studies systems of increasingly greater complexity, a "new science" will "emerge." But there is nothing "emergent" about the concepts of a science like meteorology that deals in complex systems; it merely requires a new layer of abstraction if the systems are to be made amenable to study. Such abstractions enable useful scientific work to be done that would otherwise be impossible.

Chaos

Suppose that you are feeling unwell and that you measure your temperature and then call your doctor. If you report a temperature of 37°C (98.6°F), you will told to stop malingering and go to work; if you report a temperature of 38°C (100.4°F), you will be told to rest in bed and drink tea; if you report a temperature of 40°C (104.0°F), you will be told to seek medical care immediately. Let us now ask the question: How would the advice change if you reported temperatures of 37.1°C, 38.1°C, 40.1°C (98.8°F, 100.6°F, 104.2°F), respectively? The answer is that the advice would be exactly the same, because, clinically, a tenth of a degree has little or no significance. If you invented a medical thermometer that—for a modest increase in price—could reliably measure the temperature of the human body to the nearest hundredth of a degree, you would have difficulty finding a mass-market for the product.

Every scientific measurement is subject to an error of measurement. It may be in the first decimal place or it may be in the tenth decimal place, but the error will always be there. The natural assumption is that, for a sufficiently precise measurement, the error is not significant. For clinical use in medicine a few significant figures is usually sufficient, while for a semiconductor fabrication plant the requirements might be stricter.

The assumption is that a small change in the value of a quantity in a physical system will cause a correspondingly small change in the state of the system. Science, especially physics, has always studied systems that are stable, in the sense that a small change in an initial condition results in a small change in the measured output (sidebar). Primarily, this was be-

---

Initial Conditions in Classical Physics

The equation $t = \sqrt{2h/g}$ gives the time of fall of an object, where h is the height in meters and $g \approx 9.81$ meters/second^2 is the acceleration at the surface of the Earth. If you ask for the time of fall of an object from a height of 100 meters, the answer is $\sqrt{200/9.81} \approx 4.51$ seconds. If by mistake you use a height of 99 meters as the initial condition, the time of fall is $\sqrt{198/9.81} \approx 4.49$ seconds. An error of one percent in the initial condition, leads to an error of $0.02/4.51 = 0.44$ percent in the answer.

cause in mathematics, *closed solutions* exist primarily for linear equations that demonstrate this behavior. Some mathematicians like Henri Poincaré looked into nonlinear equations, but before the advent of computers, performing extensive calculations to obtain numerical solutions of nonlinear equations was infeasible.

In 1962, meteorologist Edward Lorenz (1917–) was using an early computer to obtain numerical solutions for a system of twelve equations that were intended to model the weather. One day, to save time when working with the slow computer, he attempted to restart a calculation from the middle, entering by hand the interim values that had been printed by the computer. Although the computer printed numbers to six decimal places, he assumed that nothing weird would happen if he entered only three digits. He was wrong; something weird happened. The solution he obtained was significantly different from the one obtained using six digits. In other words, the system was very sensitive to the initial conditions. It has since been discovered that this unstable behavior can occur in the solution of even very simple equations. The sidebar on the next page shows an example of a simple, deterministic equation that displays unstable behavior.

Chaos is the name given to the study of instability in nonlinear systems. (The term was given to this field by mathematician James Yorke [1941–].) As with the term *natural selection*, this is another terribly unfortunate case where the narrow technical meaning of a term is overlayed by the totally inappropriate connotation of its colloquial meaning. There is nothing chaotic—in the sense of messy or nondeterministic—about the mathematics of chaos! It is the study of a certain class of otherwise mundane equations that simply happen to have the property that the solutions of the equations are extremely sensitive to the initial conditions.

From a historical point of view, it is not surprising that the study of chaos arose during research in meteorology. Any science that studies the flow of fluids like water or air runs into intractable mathematical difficulties. The *Navier-Stokes Equations* describing fluid flow were formulated in the nineteenth century, but have remained unsolved ever since. The solution of these equations is one of seven *Millenium Problems* in mathematics, for whose solution the Clay Mathematical Institute offers a $1 million prize.[1]

Chaos is often called a "new science."[2] The scientific and mathematical research on chaos does have some of the characteristics of Thomas

---

The Logistic Recurrence Equation

This equation is used to model the size of a population over time as a fraction x of the maximum size that the available resources can sustain. From an initial value x_0, we can compute the values of $x_1, x_2, x_3, \ldots$, defined by the recurrence equation $x_{n+1} = a \cdot x_n \cdot (1 - x_n)$. The population in the next generation is proportional to its size in the current generation and to the available ecological "space," given by a. For $a = 2.5$ and $x_0 = 0.500000$, after one hundred generations, $x_{100} = 0.600000$. If $x_0 = 0.500001$ or $x_0 = 0.499999$, x_{100} is still equal to 0.600000, so this model is not sensitive to the initial condition. However, if $a = 3.9$, the results are $x_{100} = 0.427729$, $x_{100} = 0.385272$ and $x_{100} = 0.818057$, respectively. The model is *chaotic* because the result changes significantly for a change of only one-thousanth of a percent in the initial conditions.

Kuhn's paradigm shifts from one normal science to another one. Edward Lorenz, who continued to investigate the mathematical phenomena that he had discovered, initially had some difficulty publishing his results. Mathematicians weren't very interested in nonlinear systems of differential equations, while meteorologists regarded his work as too mathematical and of little applicability to meteorology.

Because of these difficulties, it took a decade or two for a research community (a paradigm, if you wish) to develop. The study of nonlinear systems is now a well-developed field of mathematics, and of great interest to scientists and even to nonscientists like economists, because many natural systems show similar sensitivity to initial conditions. Now that the dust has settled, chaos theory is accepted as a separate and respectable field of mathematics that has applications in science.

On the one hand, the message of chaos is optimistic, because it offers new mathematical tools to study natural phenomena, but on the other hand, its message is pessimistic because it shows that a better understanding of natural phenomena, especially, an improvement in our ability to predict their behavior, is not just a matter of more precise measurements and larger and faster computers. Since the ability to *predict* is an essential element of our definition of scientific theory, chaos may be a limiting factor in the ability of science to explain and predict, because the behavior of many natural systems is inherently unstable.

"[C]haos was the end of the reductionist program in science."[3] People who would ascribe a mystical or holistic nature to chaos theory that raises it above the mundane reductionist tendencies of mathematical physics are likely to be disappointed. Chaos shows that simple systems have complex behavior and that the mathematics describing complex behavior is similar across many fields. What does this have to do with reductionism? If you didn't like studying differential equations in college, you won't find any salvation in the study of chaos. Chaos involves the development of new mathematical techniques and their application to natural phenomena, but that is what scientists have been doing since Newton.

* * *

We have defined a scientific theory to include a mechanism that explains phenomena. While it is somewhat embarrassing that the theory of gravitation does not have a demonstratable mechanism, in general, most theories can be explained by reducing them to an underlying theory that provides the mechanism. Otherwise, our theory may just be a lucky correlation. Even if we could correlate black cats with bad luck, to rise above a superstition, we want to know what it is about "blackness" or "catness" that causes bad luck, so that we can understand why black dogs or white cats do not bring bad luck.

There is one group of scientists who do hold a somewhat special position in the hierarchy of reductions, namely, the physicists who are researching what at any one time seems to be the lowest level in nature. Once the limits of reduction were atoms, then the protons, neutrons, and electrons of which they are composed, then the quarks and gluons that form the neutrons and protons. Physicists are currently attempting to reduce these particles to lower-level entities, for example, in the theory of *superstrings* (see page 203). Of course, even if we do finally achieve the ultimate reduction, and we find out what matter really "is," this is not likely to significantlly affect the rest of science. Chemists will still study reactions in terms of the properties of molecules, meteorologists will still use concepts like cold fronts, molecular biologists will still use single letters to stand for large molecules, physicians will still measure your blood pressure, and shepherds will still know the difference between goats and sheep.

LINUS PAULING: LINKING PHYSICS, CHEMISTRY, AND BIOLOGY

I plead guilty to ignoring the science of chemistry in this book. Chemistry can be placed somewhere between physics and biology. On the one hand, from the point of view of the philosophy of science, it does not deal with the fundamental particles and forces from which the universe is made and thus provokes less philosophical debate than physics with its nonintuitive theories of quantum mechanics and relativity. On the other hand, it is less susceptible to the theological and social controversy that arises in biology because we see ourselves as biological entities.

Nevertheless, chemistry is probably more important to our daily lives than either of the other two sciences. Every headache pill you down, every drop of gasoline your car guzzles, and almost every stitch of clothing that drapes your body was produced by chemical processes that have only been understood for the past century or two. Linus Pauling (1901–1994) was the leading chemist of the twentieth century and was responsible for connecting chemistry with both physics and biology.

Linus Pauling spent most of his career at the California Institute of Technology, first as a graduate student and then as a faculty member. His initial scientific work was in the field of x-ray crystallography that is used to study the structure of materials. In 1926, he visited Germany at the height of the development of the theory of quantum mechanics. Pauling used the new theory to explain the fundamental issue of chemistry: how atoms bond to each other to form molecules.

We all learned in high-school chemistry that a sodium atom loses its outer electron to form a positive ion, while a chlorine atom receives an electron to form a negative ion, and the two ions attract to form sodium chloride, salt. This type of ionic bond is rather simple to understand, but Pauling was able to describe the behavior of electrons that gives rise to other forms of bonds: the covalent bond and a new one he discovered called resonance. This enabled Pauling to explain the ring structure of the benzene molecule, original proposed in 1865 by Friedrich Kekulé (1829–1896). Pauling was awarded the Nobel Prize for Chemistry in 1954.

Pauling's theory placed chemistry on a firm theoretical basis, enabling chemists to predict properties of materials. While Pauling's work seemingly reduces chemistry to physics, in practice, the physical theory can go only so far, and chemistry is a highly useful level of abstraction with its own concepts, theories and techniques.

Pauling then turned his attention to biochemistry, elucidating the structure and functioning of the hemoglobin molecule, and explaining the mechanism of sickle-cell anemia. He was the first to propose a helix structure for proteins, and he conjectured that the DNA molecule would have three strands. There is speculation that he might have been able to deduce the double-helix structure of DNA before Francis Crick (1916–2004) and James Watson (1928–) if he had been allowed to attend a conference in England in 1952 where Rosalind Franklin (1920–1958) showed her x-ray photographs of DNA.

After World War II, Linus Pauling was active in protests against nuclear weapons testing in the atmosphere. His political activism led eventually to his resignation from Caltech, and the US government denied him a passport for several years. Activism by Pauling and other scientists was instrumental in bringing about the Limited Test Ban Treaty of 1963. Linus Pauling won the Nobel Prize for Peace in 1962.

Linus Pauling's great strength was his ability to examine a problem and to intuitively produce a solution that only later would be verified by detailed calculation and experimentation. But this led to one of his more controversial claims: that large doses of vitamin C can preserve health and combat disease. Here we see an example of the influence of social factors on science. It is true that Pauling's stature enabled him to obtain research funding that other scientists may not have been able to obtain, but the social influence did not lead to his claims for megavitamin therapy being immediately accepted, and they have not held up under further scientific investigation. The latest research from the National Institutes of Health suggests a daily intake of about 200 mg, one-tenth of Pauling's recommendation of 2000 mg that is now considered potentially dangerous.[4] The Linus Pauling Institute (at Oregon State University) basically concurs with this recommendation.[5] This episode shows that even the theories of Nobel Prize winners are subject to the scientific processes of confirmation and refutation by empirical evidence.

Linus Pauling continued to conduct active scientific research until his death at age 93.

10 Statistics: As Reliable as a Casino's Profit

Casinos and meteorology

A meteorologist has a cushy job. Suppose that on tonight's news he predicts that there is a 50 percent chance of rain tomorrow. If it rains, he was right; if it doesn't he was also right! Heads I win, tails you lose. What could possibly be scientific about a prediction like that?

Before analyzing the scientific status of statistical predictions, let us pose another question: Would you rather be a gambler or the owner of a casino? Believe it or not, a casino takes a relatively small percentage of the bets. For example, in roulette the chance that the casino will win any given round is only 5.26 percent assuming that for every bet by one gambler, there is an offsetting bet by another (sidebar). From his share of the winnings, the owner of the casino must pay off his mortgage, meet his large payroll, and cover expenses like electricity. Nevertheless, owning a casino is *not* a risky business, provided that you attract enough gamblers, preferably those willing to bet often and in large amounts.

Let us assume that the casino owner is not cheating (he doesn't have to) and that he is using a wheel manufactured to exacting standards so that each number is equally likely to appear on each spin. The mathematical theory of probability predicts that after playing a very large number of these events, the casino will win 5.26 percent of all bets. Therefore, the success of a casino depends not on luck, but on its ability to attract gamblers so that the number of bets will be very large. The physical layout and operations of a casino are designed in accordance with psychological principles that have been finely honed through years of experience. Casino owners are perfectly happy if (occasionally) a gambler wins a lot of money, because the publicity can only bring in more gamblers. Furthermore, it is usually not hard to convince the winner to continue gambling and lose what he has won.

Roulette

The roulette wheel has 38 pockets labeled 1 to 36, 0 and 00. You can bet on a specific number or that the result will be even or odd, or high (19–36) or low (1–18). You win $35 for a successful $1 bet on a single number that occurs with a probability of $1/38 = 2.6\%$, and $1 for the high/low or even/odd bets that occur with a probability of $18/38 = 47.4\%$. The probability that the ball falls into the pockets numbered 0 or 00, a win for the casino, is $2/38 = 5.26\%$.

Your *expectation* is the average amount of money per bet that you can expect to receive in a long series of bets. For an even bet, you have an $18/38 = 0.474$ chance of winning $1 and a $20/38 = 0.526$ chance of losing $1, so the expectation is $(1 * 0.474) + (-1 * 0.526) = -.052$, that is, an average loss of about 5 cents per $1 that you bet. If you start with a stake of $10,000 and bet $100 once a minute, you can expect to be wiped out in $10,000/5 = 2,000$ minutes, which is 1.3 days.

Now we can understand what it means for a meteorologist to predict that there is a fifty percent chance of rain. Keep a record of his predictions for many years and compare it with the weather that actually occurred. Suppose that on 100 days he predicted a 50 percent chance of rain. If it rained between 46 and 54 of those days, then his forecast was highly accurate. The tools of statistics can be used to take the entire record of predictions and subsequent observations of the weather, and use it to compute the quality of his forecasting. This would be expressed in a statement of the following form: There is only a 4 percent chance that the accuracy of his predictions was the result of chance rather than his skill at forecasting.

Let us now consider how we should react when we hear that the weather forecast includes a 50 percent chance of rain? There is a difference between the probability of an event and *risk tolerance*. The former is an objective prediction (though subject to error, of course), while the latter is a subjective judgment of the consequences of each outcome, of your willingness to tolerate those consequences and of the actions that you will take to prepare for each outcome. Ask yourself the question: Do I take an umbrella with me tomorrow? The answer will be based on your risk tolerance, which varies from person to person, and even an individual may modify his risk tolerance according to the situation. If you are going to the swimming pool, you might "chance it" and leave the umbrella at home, because you don't really care if your hair gets wet, but if you are going to get your hair cut and styled for your wedding, you will almost certainly take an umbrella. In both cases, you accept the forecast that there is a 50 percent chance of rain, but your risk tolerance changes.

Correlation in medical science

Medical science is the most familiar of the statistical sciences. Let us analyze one result from medical science: Smoking is hazardous to your health. What exactly does this mean? And how can the result be reconciled with counterexamples such as someone's grandmother living to be one-hundred years old even though she smoked like a chimney.

The term "hazardous" can be given a quantitative meaning. The Web site of the American Cancer Society supplies mountains of data.[1] One item is the *relative mortality risk* for smokers to die of lung cancer, which is 22.4 for males and 11.9 for females.[2] These numbers measure the relationship between the behavior of smoking and the chance of dying from lung cancer. For every one male nonsmoker who dies of lung cancer, 22.4 male smokers die of lung cancer, and for every nonsmoking female victim there are 11.9 female victims who smoked. How many people actually die of lung cancer? These data are also tabulated. In the United States, the *incidence* of death from lung cancer is 72.6 for males and 43.5 for females. This means that 72.6 of every 100,000 males will die in any year from lung cancer, and similarly 43.5 of every 100,000 females. Combining this figure with the relative mortality risk, we see that about 3 male nonsmokers out of every 100,000 people die every year from lung cancer as opposed to about 69 male smokers. Some people get a bad deal, abstaining from smoking and yet succumbing to the disease, while other people like your friend's grandmother can get away with a lifetime of smoking. Reports of particular cases are called *anecdotal* and are of limited importance. At most they can point the way for further investigation. What is special about the genes or the environment of that particular lady?

These data are "raw," in the sense that they are just data that have been collected and tabulated, but not interpreted. All they say is that lung cancer is *positively correlated* with smoking. This means that if you walk into a hospital and ask a random lung cancer patient if he smokes, the answer will quite probably be yes. But correlation alone is not *causality*. Prostate cancer is positively correlated with male gender, while ovarian cancer is positively correlated with female gender, yet no one would claim that being male *causes* prostate cancer and being female *causes* ovarian cancer.

Drawing conclusions from correlation

Before analyzing causality in the next section, let us discuss here conclusions that can be drawn from correlation alone. It turns out that data about mortality rates are highly accurate, in fact, so accurate that you could bet on them if there were a casino willing to take such wagers. Such casinos are not hard to find; they are called life insurance companies. You place a relatively small bet called a premium and if you "win," they pay you (or rather your family) a large sum of money called a benefit. For example (using arbitrary round numbers), let us assume that paying a $100 yearly premium will entitle the family of the policy owner to receive a benefit of $100,000 in case he dies of lung cancer, and let us further suppose that the insurance company convinces one hundred thousand men to buy this policy. The company receives $100 × 100,000 = $10 million and pays out 72.6 × $100,000 = $7.26 million (more or less), leaving them with a tidy profit of about $3.84 million, minus costs and the inevitable taxes. Quite a good business! However, the business is only as good as the statistics, because if for some reason 120 of the men who paid the premium die of lung cancer, the company will pay out $12 million for a loss of $2 million.

The insurance industry exists only because certain important risks are *quantifiable*, and moreover, reliably quantifiable. Insurance companies will refuse to cover certain risks—like damage caused by a collision of an asteroid with the Earth—not because such risks are necessarily rare, but because they are unable to reliably quantify the risk. Insurance companies do in fact accept the statistical data that smoking is hazardous and they are willing to back this belief up by betting money. Based upon the immense difference in the mortality rates of smokers and nonsmokers, they offer differential premiums to these two groups of people. To continue our hypothetical example, suppose that the company insures one hundred thousand people, all of whom are nonsmokers. They can still make a decent profit by asking for much smaller premiums of say $10 each, since they can expect to pay out only about $300,000 of the $1 million they take in premiums. The premium is quite a small sum and many people would be willing and able to make such a bet; if a million people bought this coverage the profits would again be tempting: $10 million taken in and only $3 million paid out. Conversely, they would not reduce the premiums for smokers because the company can expect to "lose" large numbers of bets.

So why is it considered responsible to take out life insurance, yet irresponsible to gamble at roulette? There are two differences. The first is that the payoff in insurance is intended to avoid specific financial catastrophes. If you die, the insurance benefit will enable your family to remain in their house and pay for your kids to go to college, while in gambling, the payoff is not connected to any particular event. Your winnings in a casino (to the extent that you have any) are not guaranteed to be concentrated only when your life savings evaporate in a stock market crash and you truly need the cash. If you could receive such a guarantee, gambling might be worthwhile.

The second, more important, difference is that your bets with the life insurance company are fixed and predictable. You do not pay them premiums hour after hour, day after day, until you have exhausted all your financial resources, and mortgaged your home and business to loan sharks. Gambling is a serious social problem, not because it causes people to lose money, but because many people are compulsive gamblers. You might be willing and able to pay $20,000 to enjoy a first-class trip to Paris for a romantic weekend to celebrate your twenty-fifth wedding anniversary. If you can afford such entertainment, there is nothing wrong with it (unless you start feeling guilty about enjoying yourself instead of contributing to a worthy charitable cause). The situation is completely analogous to dropping $20,000 in a casino on a *one-time* preplanned binge; it is just entertainment. But of course casinos would not survive long if gamblers bet only on very special occasions and set strict limits on their cumulative losses.

When does correlation become causality?

The step from correlation to *causality* is fraught with danger for the same reasons that make any scientific inference difficult. You can never be absolutely sure that you are not seeing an illusion or ignoring a significant but unknown factor. It is not unusual, for example, to see a report of medical research claiming that eating something is positively or negatively correlated with contracting a disease and then a few years later to read of research claiming the exact opposite. The human body is a complex structure, where billions of cells interact through physical and chemical mechanisms directed by tens of thousand of genes. For any particular individual, there is no way to accurately predict the result of some treatment, which is why warning labels of medications contain long lists of dangerous and sundry side effects that may occur, even if only rarely. We are back in the casino—you have to bet that the potential benefits of a treatment are much more likely to occur than the possible dangers, and, more importantly, you have to accept the risks involved.

Nevertheless, at some point, we begin to see causality rather than mere correlation. We can say that a factor *causes* a phenomenon when:

- There is a high positive correlation between the factor and the outcome.

- These correlations occur reliably in repeated experiments using a variety of techniques.

- There is a mechanism to explain the correlation.

Under this definition, we can definitely say that touching a hot stove *causes* burns. There is a high positive correlation between touching a hot stove and getting burned. The correlation is reliably obtained in repeated experiments (though you may prefer not to carry out repeated experiments yourself, but merely observe or read about other cases). Finally, from physics and chemistry, we know that raising the temperature of tissue increases molecular disorder, breaking chemical bonds, and ultimately producing the biological changes that are observed as burns. You do not have to understand physics and chemistry in order to modify your behavior and avoid touching hot stoves. Even in the absence of knowledge of the mechanism, knowledge of the reliable correlation justifies your actions. Unfortunately, few situations are as clear-cut as burns caused by a hot stove. Some people don't smoke and die of lung cancer, while others do smoke and live to be a hundred. How can we establish causality in this case? There are three basic means: experimentation, epidemiology, and research on potential mechanisms.

Medical experimentation

In its initial stages, medical research can be carried out in the lab. You can apply an antibiotic directly on a colony of bacteria in a dish and see if they die. However, this does not provide evidence that the antibiotic will kill bacteria within the human body, or that it will do so without severe side effects. Once there is evidence that a procedure or a drug is likely to be safe and effective, a *clinical trial* can be carried out in which a control group is given the traditional drug and a treatment group is given the new drug. If the results for the treatment group are better than, or at least as good as, the results for the control group, then there is evidence that the new drug can be used instead of the traditional drug. There are statistical means for specifying precisely what is meant by "better than, or at least as good as."

Great care must be taken to ensure that the experiment really confirms or falsifies what is intended. In particular, both the traditional and the new drug must be given in as similar a regimen as possible; if there is no traditional drug, an inert placebo is given to simulate the regimen of a drug treatment. You do not approach a patient and say, "Did this revolutionary new drug relieve your heartburn?" The patient will be reluctant to disappoint the young doctor desperate for a journal publication and to admit that the revolutionary new drug did not work on him. Even if the outcome of the treatment can be objectively measured like the glucose level in blood, this experimental procedure is still followed because the behavior of the patient may influence blood chemistry.

Medical experiments should be conducted *double-blind*, meaning that even the clinician administering the treatment should not know which patient receives the traditional drug and which the new one. This comes not so much from suspicion at the possibility of outright fraud, which occurs though relatively rarely, but from the possibility of a subconscious effect on the interpretation of the findings. "So, you're taking the revolutionary new drug and all you had was an occasional twinge of heartburn; well, a few twinges are hardly significant and can be ignored."

The design of clinical experiments is difficult because of the ethical dimensions involved. This subject is much too complex to treat here, except to note that ethical considerations often make clinical trials much less satisfactory than you would want from a purely scientific perspective. For example, if a new treatment seems to be causing severe side effects, the experiment might have to be discontinued, even though it might have turned out that the side effects are temporary and worth the life-saving improvement that the treatment brings. Conversely, if a new treatment seems to be especially effective, it becomes unethical to continue the experiment and to deny the new treatment to the control group, even though from a scientific point of view it would have been better to carry out the full extent of the experiment.

Even when due care is taken, a single medical experiment can provide only partial evidence for causality. The reason is that conditions cannot be fully controlled. The researcher can match the age, gender, and ethnic makeup of the members of the control and treatment groups, but she certainly cannot match their detailed genetic structures and detailed lifestyles.

Another type of medical experiment utilizes analogs, such as animals. The anatomy and physiology of all mammals are quite similar and they provide analogs for humans. Furthermore, animals like mice and rats can be bred to be genetically identical and can be raised in identical conditions. The results of an experiment on mice are quite reliable and easy to repeat, but mice are still just analogs and there is always the possibility that the results will not be applicable to humans.

Epidemiology

Epidemiology is the study of disease by statistical analysis of the occurrence of the disease and the associated environmental or genetic factors. It has a long history, from elementary observations that contact with the victim of a plague can be infectious, to Edward Jenner's observation in 1796 that milkmaids infected with cowpox were not susceptible to smallpox. The scientific study of epidemiology began in the nineteenth century with the work of John Snow (1813–1858) in London and Ignaz Semmelweis (1818–1865) in Vienna. Both these physicians used quantitative methods to obtain reliable correlations between factors and disease. These correlations in themselves were sufficient justification to modify behavior, but a full explanation of causality had to wait for many years until the mechanism of the diseases—infection by bacteria—was discovered.

The pioneer of modern epidemiology was John Snow who studied outbreaks of cholera in mid-nineteenth-century London. Through painstaking fieldwork, Snow was able to correlate cases of the disease with the source of the victim's drinking water. (The biographical vignette of John Snow at the end of this chapter tells the story in detail.)

In the 1840s, Ignaz Semmelweis, a doctor in the Vienna General Hospital, noticed an extremely high rate of mortality from puerperal infection in one obstetrics ward. He investigated many possible correlations and found only one factor that highly correlated with mortality: doctors visiting that particular ward directly after performing autopsies. When he and his students cleaned their hands thoroughly with a disinfectant before treating patients, the mortality rate returned to normal levels. Only in 1890 did Louis Pasteur show that the mechanism of these infections was *Streptococcus* bacteria.

The problems facing Snow and Semmelweis were relatively simple because the rate at which a disease manifests itself following bacterial infection is rapid. Drink infected water and within a day or so you will be ill with cholera. Shut off the supply of infected water and people will stop coming down with the disease. Many genetic diseases such as cystic fibrosis and hemophilia are similarly clear-cut. If you have the appropriate genes, you have the disease, and if not, you don't. However, the progress of medical science during the last century has clarified the mechanisms of most infectious diseases, as well as those that are definitively caused by genetic changes. Interest is now focused on diseases like cancer and heart disease that take years or decades to develop. If a modern-day Dr. Snow closed all pizza parlors and fish-and-chip shops in London for a week, the mortality rate from heart disease would not be immediately affected, nor would the mortality rate from lung cancer in Vienna be reduced if a modern-day Dr. Semmelweis forced all residents to refrain from smoking for a month.

To establish the existence of positive correlations between smoking and cancer, or between fatty foods and heart disease, it is necessary to carry out long-term epidemiological investigations, either forward or backward. To perform a forward investigation, a large sample of people is chosen and followed through years or decades, using questionnaires, tests, and interviews to elicit data about their environment and habits. Comparing this data with their medical condition enables the researcher to demonstrate correlations between risk factors and disease. The main difficulty is to ensure cooperation over such a large period of time.

Alternatively, you can start with a large sample of sick people (or even just their medical records) and work backward by asking them or their relatives about their habits. Their habits can then be compared with a similar sample of people who are not afflicted with the disease. Here the sample is likely to be more representative, though the accuracy of the data may be poor due to forgetfulness.

With either method, statistical techniques must be used to ensure that factors are not confounded. Perhaps the same genetic predisposition to suffer from stress causes a predisposition to smoke, so that if lung cancer were caused by stress, we would still see a positive correlation with smoking. Since humans are so complex both physically and psychologically, any epidemiological study by itself can never be fully conclusive. There will always be some bias in the selection of the sample and some confounding factors that cannot be fully accounted for. Nevertheless, careful studies provide powerful evidence on the road toward demonstrating causality.

The statistics published by the American Cancer Society show that the mortality rates for lung cancer in the state of Utah are only 28.6 for males and 14.0 for females, while the rates for the state of Kentucky are 99.4 and 44.2, respectively. The obvious explanation is that many residents of Utah belong to a religion (the Church of Jesus Christ of the Latter-Day Saints, also known as the Mormons) that requires their members to abstain from smoking, while Kentucky is a state in which tobacco growing is an important industry, so smoking is much more likely to be socially acceptable. But Latter-Day Saints also refrain from drinking alcohol, which is produced and consumed in Kentucky, so it is possible that whiskey causes lung cancer. This could be checked by examining the mortality rates for devout Muslims, who refrain from drinking alcohol but not from smoking. These are rather simple examples of the confounding factors that must be considered when designing and performing epidemiological studies. The genetic makeup, environment, and lifestyle of one person will differ from those of another person, but if appropriate data is collected, advanced statistical techniques can calculate the contribution of each factor to the outcome.

Mechanism in medical science

Because of the lingering doubts that may accompany epidemiological studies, a demonstration of causality must ultimately include the identification of a pathogenic (disease-causing) mechanism. It would not be wrong to say that this is the main difference between modern and premodern medicine, or between modern medicine and pseudosciences like homeopathy. Before the advent of modern medical science, treatments were discovered by trial-and-error. Sometimes they were effective like the use of quinine to treat malaria, but at other times treatments like blood-letting caused more harm than the disease itself. The first mechanisms discovered were microorganisms like bacteria and parasites, and recently medical science has identified other mechanisms such as genetic defects and chemicals. Frequently, a combination of mechanisms is needed to cause a disease, for example, genetic susceptibility combined with exposure to chemicals.

The mechanisms that cause lung cancer have been thoroughly investigated and include changes in lung tissue that can be associated with carcinogenic agents in tobacco, so on this question, medical science has reached far beyond the stage of mere correlation.

It is not necessary to wait for the elucidation of mechanism to change your behavior; even a set of high-quality studies demonstrating correlation should be sufficient. After all, people avoided malarial swamps long before the mechanism of a pathogenic parasite and its mosquito vector was discovered.

Making decisions in statistical sciences

In a statistical science like medical science, it is the preponderance of evidence from many sources, rather than one decisive experiment, that justify a scientific conclusion. The preponderance of evidence—epidemiological and biological—clearly justifies the statement that smoking *causes* cancer. For other factors and other diseases, the evidence is there, but pieces of the biological and epidemiological puzzle are lacking. Furthermore, not every laboratory finding is necessarily clinically relevant, and not every clinically feasible procedure is necessarily economically feasible.

In the end, your decisions must be based on your risk tolerance. While few of us have the time and knowledge to fully evaluate every scientific report, most of us can come to a reasonable assessment of risk based on high-quality reports of scientific results published for the general public. This, combined with an assessment of the consequences, will enable us to make reasonable decisions. Here are some examples of scientific dilemmas and issues of risk tolerance.

Mad cow disease: Mad cow disease is the media-friendly name for *bovine spongiform encephalopathy (BSE)*, a neurological disease affecting cattle. BSE is similar, both clinically and in the morphological changes in brain tissue, to a disease affecting humans called *variant Creutzfeldt-Jakob*

Disease (vCJD). Scientists suspect that vCJD is caused by the same mechanism as BSE, because many cases occurred in the United Kingdom which had a severe outbreak of BSE. Nevertheless, the epidemiological evidence is incomplete because the number of patients is extremely small, fewer than two hundred in a decade. For any victim who might have contracted vCJD from eating a hamburger, there are dozens of neighbors who ate the same hamburgers and did not contract the disease.

The claim that vCJD is caused by eating meat from BSE-infected animals was strengthened with the discovery of *prion proteins* as a possible mechanism for transmitting the disease. We do not have a full understanding of the epidemiology and biology of the danger to humans from mad cow disease, but the consequences are so devastating that some level of prudence is advised. Here is a list of prudent choices, arranged in order of increased risk tolerance: don't eat meat at all, don't eat beef, don't eat beef in the UK, don't eat bone and nervous system products from cattle in the UK, don't eat such products until after the source has been certified as disease-free. Since each individual assesses risk differently and has a different level of tolerance for risk, each of us must make our own decision as to what constitutes prudent behavior.

Ketchup prevents cancer: We are daily bombarded with nutritional advice like "ketchup prevents cancer." Sometimes this comes from an experimental result describing a mechanism, which may or may not be backed up by clinical and epidemiological evidence. At other times, a single epidemiological study seems to point in a certain direction but the evidence is far from conclusive. Scientific research has shown that chemicals called oxidants can be shown to cause tissue damage that can lead to cancer, and that antioxidants such as *lycopenes* found in tomatoes can help prevent oxidation.

But there is a long way to go from this mechanism to the clinical advice that ketchup can prevent cancer. Epidemiological studies must be performed to demonstrate that—all other factors being equal—people who consume more ketchup and other tomato products have a lower mortality rate from cancer. Such a study is practically impossible to carry out, because eating tomatoes is not an activity as clear-cut as smoking cigarettes. Can you retroactively quantify your consumption of ketchup and tomato paste?! Furthermore, there are an inordinate number of potentially confounding factors that must be controlled for; in particular, we must control for exposure to both oxidants and antioxidants in many kinds of food.

Screening for colon cancer: Colon cancer is a serious disease that can be cured if diagnosed and treated at an early stage. There are many diagnostic procedures that can be used: stool samples, imaging techniques (x-rays, CT scans) and endoscopic procedures (sigmoidoscopy, colonoscopy). There is no question that if suspicious symptoms are present, the physician should prescribe the appropriate procedures until a definitive diagnosis is made. However, this is a fairly typically case in modern medicine where diagnostic procedures can also be used to *screen* populations of symptom-free individuals. in many cases detecting pathologies before they manifest themselves clinically.

Each procedure has associated with it a cost composed of: (a) the financial cost of the procedure, (b) the patient's inconvenience, discomfort and anxiety, (c) the time of highly trained medical specialists, (d) the damage that may be caused by complications of the procedures themselves and their treatment, and, above all, (e) the cost engendered by potential inaccuracies in the procedures. An inaccuracy can be a *false positive*, where the procedure says you have it, but you don't, leading to further unnecessary tests and heightened anxiety, or a *false negative*, where the procedure says you don't have it, but you do. False negatives are particularly pernicious because they lead to false complacency. When you finally start having symptoms, you will dismiss them until it is too late because the procedure had shown that you were free of disease. Generally, the more complex procedures are more accurate, but they are also more expensive, have more complications, and are more difficult both for the patient and for the physician. This leads to lower levels of patient compliance and a greater waste of time of highly trained physicians who could be better employed diagnosing and treating people with symptoms associated with the disease.

Health organizations publish guidelines trying to balance the trade-off involved, for example, by trying to identify populations who are at greater risk and recommending that they be screened more aggressively than the general population. But there can never be an unambiguous decision that will be totally correct, only a reasonable decision taken in the presence of statistical uncertainty.

* * *

There is a term *physics-envy* that is used to describe the overmathematization of fields of study that attempt to achieve the same level of accuracy and precision that is possible in physics research. Scientific theories must be able to precisely and accurately predict natural phenomena, but a certain amount of flexibility is necessary when dealing with complex systems like living organisms. The predictions may only be statistical, but that does not make them any less valid as scientific theories, as is sometimes claimed by purveyors of pseudoscientific medical treatments. It does mean that the scientist must invest a lot of effort in designing and analyzing experiments to ensure that they are statistically sound. More importantly, it requires more effort on the part of the consumers of research results to interpret them and to devise reasonable responses.

In principle, medical decisions are no different from the decision whether or not to take an umbrella when the forecast calls for a fifty percent chance of rain, but their importance requires a more rational approach. We do not live in a world in which our future can be forecast with perfect accuracy. We have to do as the croupier says: *Faites vos jeux, mesdames et monsieurs.* Certainly though, you should know the odds and the possible consequences, for once you take a drag on that cigarette or bite into that hamburger, *rien ne va plus.*

JOHN SNOW: SAVIOR OF LONDON

London is one of my favorite cities. I particularly like to browse the labyrinth of stacks in Foyles bookshop on Charing Cross Road or to marvel at the Rosetta stone and other historical gems in the British museum. It is hard to believe that just 150 years ago during the first decades of the Victorian era, London was filthy. Raw sewage was dumped in the Thames River, and drinking water was drawn directly from the Thames or from shallow wells, and drunk with little or no filtering. Paradoxically, the Sanitary Reform Movement had contributed to the problem by constructing a sewer system to replace cesspools; instead of limiting contamination to local areas, the sewer system ensured that contaminants were widely distributed through the water of the Thames.

Needless to say, disease was rampant, including repeated epidemics of cholera. Cholera is a terrible disease caused by *Vibrio cholerae* bacteria. After a short incubation period, often only one day, it causes massive diarrhea. Untreated, some patients recover, but many die within a day or so from dehydration and electrolyte imbalance. If it is caught in time, treatment is easy and quite effective, consisting of antibiotics to kill the bacteria and the replacement of lost fluids and salts, either orally or intravenously. Before the discovery of bacteria, the cause of cholera was unknown. Most scientists believed that the disease was caused by *miasmas*, mysterious emanations wafting through the air.

In a pioneering epidemiological study, Dr. John Snow (1813–1858) studied outbreaks of cholera in London in the mid-nineteenth century. Snow painstakingly mapped out cases of cholera and interviewed the victims' families in order to obtain as much information as possible about their contacts with other victims, their professions, and their habits, in particular, the source of their water. Snow concluded that cholera was transmitted by a water-borne agent. Sewage and water from the laundry of soiled bed linen leaked into wells or was dumped into the Thames river and from there it got into the drinking water. In addition, caretakers ministering to the sick prepared food without first washing their hands.

Here are some examples of Snow's results:

- A sudden massive outbreak of cholera hit the Soho area in London on August 31, 1854. Within a couple of weeks, 500 people had died, mostly within a small area. Snow was able to show a high correlation between contracting the disease and drinking water from a pump at the corner of Broad and Cambridge streets (since renamed Broadwick and Lexington streets, not too far from Foyles and the British Museum).[3] The photograph on the next page shows a pump erected in Soho as a monument to John Snow. The following page contains a closeup of the plaque at the foot of the pump.[4]

- Conversely, neighbors who did not drink from the pump contracted the disease at extremely low rates. These included inmates in a prison that had its own well and brewery workers who imbibed beer instead of water.

- Strange as it may seem today, the provision of water was a competitive business, with water pipes belonging to rival companies supplying the same street. Snow calculated that during the 1849 epidemic, the incidence of cholera in customers of both the Lambeth and the Southwark & Vauxhall water companies was similar. By the epidemic of 1853, the Lambeth company had moved its water intake to a source above most of the sewage outlets, so that its customers suffered a much lower incidence of cholera, and the cases that did occur were attributed by Snow to cross leakage between the pipes.

- Snow went so far as to infer that cholera abated during the winter in England, because—needless to say—the English drank mostly tea in the winter and boiling the water killed the pathogenic agent. In Scotland, however, people drank unboiled water all year long, so there was no significant difference in the incidence of cholera between the winter and the summer.

Snow's book is a model of scientific investigation and is quite readable even today. He marshals his data, interprets them, and refutes alternative explanations. The evidence is completely convincing and all that was lacking was a mechanism. Coincidentally, in the same year that the cholera epidemic raged in London, Filippo Pacini (1812–1883) discovered that *Vibrio cholerae* bacteria were the mechanism that gave rise to cholera. His work was ignored and it was not until 1884 that Robert Koch (1883–1910) rediscovered the cholera bacillus and it was gradually accepted as the mechanism that caused the disease.

John Snow was an accomplished scientist and physician specializing in physiology. In addition to his work in epidemiology, he was one of

THE SOHO CHOLERA EPIDEMIC

DR. JOHN SNOW (1813-1858) A NOTED ANAESTHETIST, LIVED NEAR THE FOCUS OF THE 1854 SOHO CHOLERA EPIDEMIC, WHICH CENTRED ON BROAD STREET, AS BROADWICK STREET WAS THEN CALLED. IN SEPTEMBER OF THAT YEAR ALONE, OVER 500 PEOPLE DIED IN SOHO FROM THE DISEASE

SNOW HAD STUDIED CHOLERA IN THE 1848-9 EPIDEMIC IN SOUTHWARK AND WANDSWORTH. HIS THEORY THAT [PO]LLUTED DRINKING WATER WAS THE [...] OF TRANSMISSION OF THE DISEASE [C]ONFIRMED WHEN HE MAPPED

CHOLERA DEATHS IN SOHO WITH THE SOURCE OF THE VICTIM'S DRINKING WATER. HE FOUND THAT THEY WERE CONCENTRATED ON THE BROAD STREET PUBLIC WATER PUMP

HIS THEORY INITIALLY MET WITH SOME DISBELIEF BUT SUCH WAS HIS CONVICTION THAT HE HAD THE PUMP HANDLE REMOVED TO PREVENT ITS FURTHER USE. SOON AFTERWARDS THE OUTBREAK ENDED

THE ORIGINAL PUMP IS BELIEVED TO HAVE BEEN SITUATED OUTSIDE THE NEARBY "SIR JOHN SNOW" PUBLIC HOUSE

THIS WATER PUMP

WAS UNVEILED BY

COUNCILLOR DAVID WEEKS

LEADER OF WESTMINSTER CITY COUNCIL

ON

20 JULY 1992

IT MARKS A PIONEERING EXAMPLE OF MEDICAL RESEARCH IN THE SERVICE OF PUBLIC HEALTH

City of Westminster

THE PLACEMENT OF THIS ARTEFACT AND ASSOCIATED ENVIRONMENTAL IMPROVEMENTS IN BROADWICK STREET HAVE BEEN GENEROUSLY SUPPORTED BY LYNTON plc the pioneers of anesthesiology, using his scientific knowledge to develop new equipment and clinical techniques. His successful administration of chloroform to Queen Victoria during her eighth delivery in 1853 was instrumental in convincing public opinion of the safety of anesthesiology.

The next time you read a media report saying that "X causes Y," please compare that report to John Snow's book. Did the researcher carry out a full epidemiological study? Did he perform clinical experiments to show that "X causes Y" in people and not just in mice? Did he rule out other possible explanations?

And the next time you feel a wave of nostalgia for Victorian England, remember Dr. Snow and the cholera epidemic.

11   Logic and Mathematics: Scientists Like It Clear and Precise

Deductive logic

The Greeks initiated the study of logic as formal laws to guide thinking. They were particularly interested in rhetoric, the art of communication, especially oral communication, and especially for the purpose of persuading another person or an assembly of people to accept one's ideas and views. Unfortunately, training in rhetoric is no longer part of the school curriculum, but deductive logic is still used informally by all scientists and mathematicians.

A *sound* deductive rule preserves logical truth, that is, if the premises are true, so is the conclusion. An example of a sound deductive rule is *modus ponens*:

> Premise 1: If there is no gas in the car, the car will not start;
> Premise 2: There is no gas in the car;
> Conclusion: The car will not start.

Another sound deductive rule is specialization of a universal premise that states that something is true for all objects:

> Premise 1: All cars have engines;
> Conclusion: My car has an engine.

However, not every deduction is sound and preserves truth, as shown by the following gem by Raymond Smullyan (1919–):

> Premise 1: Some cars rattle;
> Premise 2: My car is some car;
> Conclusion: My car rattles.[1]

It is particularly difficult to formulate sound deductive rules for statements that use quantifiers, which are statements qualified by "for all" or "some."

Why My Car Doesn't Rattle

Premise 1, "some cars rattle," is formally expressed by "there is some object, which has the property of being a car and also the property that it rattles." Premise 2, "my car is some car" is "the specific object *my car* has the property of being a car." You may not conclude that "*my car* rattles," because while the first premise states that there are objects which are both cars and rattle, there may be lots of cars that don't rattle and lots of things that rattle even though they are not cars.

Not until the nineteenth century was the first deductive system that dealt successfully with quantifiers proposed by Gottlob Frege (1848–1925), and logic was transformed from a branch of philosophy and rhetoric to a branch of mathematics. The goal of the first mathematical logicians, in particular David Hilbert and Bertrand Russell (1872–1970), was to base mathematics upon logic, which was looked upon as a self-evident. This goal was shown to be unachievable by Kurt Gödel (1906–1978) and Alonzo Church (1903–1995) in the 1920s and 1930s.

Nevertheless, deductive logic is an essential part of scientific reasoning because it enables specific predictions to be derived from the universal laws of a scientific theory. The use of deduction is second nature and usually causes no problems, provided that we remember that it is limited to showing logical consequence: the truth of a conclusion follows from the truth of the premises. Logic can say nothing about the truth of the premises. Furthermore, most scientific theories contain idealizations that never occur in practice; for example, the theory of gravitation gives a law (page 17) for the force existing between two objects; in practice, every object in the universe affects every other object and there is no way of isolating just *two* objects. Therefore, scientific debate revolves not around disagreement about deductive logic, but around disagreements concerning the premises of a deduction: Are the premises true? Are the idealizations valid in a particular situation?

Logical fallacies

Though the use of deductive logic is second nature to scientists, unsound deduction is a hallmark of much of modern communication. Unsound rules of deduction are called *logical fallacies.* You can find lists of fallacies on some Web sites, as well as in Carl Sagan's book *The Demon-Haunted World.*[2] Some of the fallacies are relevant to any form of public rhetoric, not just to science. For example, you are using the logical fallacy called *argumentum ad hominem,* when you attack the person rather than his views. Here we summarize several fallacies that are of particular importance to science because they are frequently used by proponents of pseudoscience.

Argumentum ad antiquitatem: "Arguing from antiquity," as exemplified by astrology, which is justified by appealing to the wisdom of the ancient Greeks.

Cum hoc ergo propter hoc; post hoc ergo propter hoc: These fallacies confuse correlation with causality, and claim a causal relationship merely because two events occur together or one after another, respectively.

Argumentum ad ignorantiam: In this fallacy, a claim is made that something is true until it is proved false. This fallacy is also a favorite of pseudoscientists. Since science cannot always supply sufficient evidence for the truth of a claim, this is taken as proof of the truth of a pseudoscientific counterclaim. For example: "This remedy has never been proved not to be effective, therefore it is effective," or, "There must be something in the claims of astrology, because no one has ever proved that the stars cannot affect humans."

Argumentum ad logicam: Claiming that a proposition is false simply because the argument is unsound. John Dalton (1766–1844), the father of modern chemistry, showed that atomic elements combine in integral proportions to form molecules. Although his conclusions were basically true, all of his premises are now known to be false![3] For example, he assumed that the sizes of different atoms are significantly different, which is not true. You would be committing the fallacy of *argumentum ad logicam* to reject Dalton's conclusion just because his reasoning was unsound.

The passes-for fallacy: What has *passed for* science has frequently turned out to be incorrect (for example, phlogiston in chemistry, ether in physics, spontaneous generation in biology); therefore, the epistemological and social basis of science (for example, objectivity, acceptance of facts, and theories based upon evidence) must be incorrect as well.[4] This fallacy is beloved by postmodernists and pseudoscientists who are eager to undermine science.

Material implication

There are some fallacies that truly deserve the name *logical* fallacies because they stem from misunderstanding the form of a deductive rule. Primarily, this is due to the use of *material implication* in logic. Consider the following two sentences:

> If Napoleon was ill on June 18, 1815, then Napoleon lost the Battle of Waterloo.

> If the Moon is made of green cheese, then Napoleon lost the Battle of Waterloo.

The first sentence is clearly a true sentence, as both "Napoleon was ill on June 18, 1815" and "Napoleon lost the Battle of Waterloo" are true statements, but surprisingly, the second sentence is also true, even though the premise "the Moon is made of green cheese" is false. The surprise comes from an incorrect attribution of causality to implication.

On June 18, 1815, Napoleon suffered from a urinary infection that caused fever and pain. He lay in bed most of the day instead of directing the battle, and this was almost certainly one factor in his loss of the battle. However, this causality is not relevant to an analysis of the implication in the sentence. To analyze the truth of an implication, we only need to ask if it *preserves* truth. But the true statement "Napoleon lost the Battle of Waterloo" is true, whether or not any set of premises is true. If the premise is true, truth is preserved; if it is false, there is nothing to preserve!

Consider now the sentence:

> If the Moon is made of green cheese, then Napoleon *won* the Battle of Waterloo.

Similarly—and this is the difficult part—since the premise is false, there is no truth to preserve, so this implication in this sentence is considered to be true. The only way that an implication can be false is if the premise is true and the conclusion is false, so that truth is *not* preserved:

> If Napoleon was ill on June 18, 1815, then Napoleon *won* the Battle of Waterloo.

The following fallacies arise from incorrectly interpreting material implication.

Denial of the antecedent: Claiming that a conclusion is false simply because the premise is. Consider again:

If the Moon is made of green cheese, then Napoleon *won* the Battle of Waterloo.

Just because the Moon is not made of green cheese does not allow us to claim that Napoleon won the battle of Waterloo.

Affirmation of the consequent: Claiming that the premise is true, just because the conclusion is true. Consider again:

If the Moon is made of green cheese, then Napoleon lost the Battle of Waterloo

We would be committing this fallacy if we used the sentence to claim that the Moon is made of green cheese. Pseudosciences employ this fallacy frequently, because it enables you to claim the truth of any premise you wish simply by choosing a true conclusion.

The logical fallacy of the affirmation of the consequent is extremely important because it shows that a scientific theory can never be proved. No matter how much evidence we amass in terms of conclusions (predictions) that turn out to be true, these conclusions do not entitle us to claim the truth of the premises. Newtonian mechanics was used to predict planetary motion in great detail and with extremely accuracy, even to the point of leading to the discovery of new planets like Neptune. Scientists assumed that Newtonian mechanics was a true theory, and then used it to predict that unexpected deviations in the orbit of Uranus must be caused by an as-yet-undiscovered planet, and they were able to predict where to look for the planet. In effect, they made the following statement:

If Newtonian mechanics is a true theory, then a new planet (with certain properties) exists.

When the planet was discovered, it was taken as "proof" of Newtonian mechanics, but logically that is an affirmation of the consequent. The conclusion "a new planet exists" is true, but that does not entitle us to claim the truth of the premise "Newtonian mechanics is a true theory." In fact, it turns that that Newtonian mechanics is incorrect and must be replaced by Einstein's theories of relativity (page 19).

Bifurcation in creationism

The arguments of creationists can be classified into two main categories. One is the argument for design discussed in chapter 2 and the other is the argument from *bifurcation*, which is the fallacy of claiming that there are only two possibilities in any situation. Henry Morris uses the fallacy of bifurcation explicitly and frequently, for example:

> There are only these two possibilities. There may be many evolutionary submodels (e.g., different evolutionary mechanisms or sequences) and various creationist submodels (e.g., different dates of creation or events of creation), but there can be only two basic models—evolution or creation.[5]

Since Morris finds serious problems with the theory of evolution, he concludes that creationism must be true. Now in a certain very limited sense, it is true that life is either a *natural* phenomenon (though not necessarily one described by the theory of evolution by natural selection) or it is a *supernatural* phenomenon. But clearly creationists are interested in a lot more than than this trivial statement "either X or not-X." The use of bifurcation reveals a lack of imagination in postulating possibilities rather than quibbles over fossil sequences and dates.

The only possibilities are *not just* the theory of evolution by natural selection and creation by a single, benevolent creator. The Raelians, for example, believe that life on Earth comes from DNA brought by extraterrestrials from outer space.[6] This, of course, just passes the question of natural vs. supernatural to the extraterrestrials, who presumably could give us the answer. Alternatively, suppose that the pagans were right and that there are many gods; in that case, plants could have been created by one god and animals by another, or, predators by one and prey by another. Or, perhaps, plants were created by the creator and animals evolved from them by natural selection; or conversely. Maybe the creator created all life forms up until the Cretaceous era, then regretted his work and sent an asteroid to destroy life, but a few furry little mammals escaped and carried on by evolution. Thinking up absurd scenarios may be just an amusing parlor game, but they do show that the argument from bifurcation is totally fallacious.

Gödel's theorems

Postmodernist criticism of science has focused on theorems proved in the 1930s by the mathematician Kurt Gödel, taking them out of their technical context and using them to make very far-reaching philosophical claims. Gödel's theorems are known at the *incompleteness* theorems, and postmodernists interpret this to mean that science and mathematics are "incomplete," from which it must follow (by the fallacy of bifurcation!) that alternate ways of obtaining knowledge about the universe are equally valid. For this reason, it is worthwhile to present an overview of these theorems so that you can appreciate their technical content, while understanding why the above philosophical interpretation is not valid.[7]

Gödel proved two theorems:

- In any consistent deductive system (comprehensive enough to define arithmetic), there will be true statements that are not provable *within* the system.

- The consistency of such a system is not provable *within* the system.

Consistency simply means that there is no statement S such that both S and not-S are provable. For example, a system would be inconsistent if both $2 + 2 = 4$ and $2 + 2 \neq 4$ were provable. It can be shown that in an inconsistent system *every* statement is provable. (For readers with an elementary knowledge of logic, the proof is given in the sidebar on the next page.) If some property characterizes *everything* then it is useless as a characterization. If everything is provable, what is the point of saying that "X is provable"? Consistency is thus a necessary characterization for any useful logical system, though it is a separate question whether it is important to actually prove consistency. Mathematicians worked for centuries without such proofs.

The dream of David Hilbert and other mathematicians specializing in the logical foundations of mathematics was to produce a deductive system that was consistent and yet able to prove all the true statements of mathematics. Gödel's tour de force proved that to be impossible, because for any deductive system there will be true statements that cannot be proved within the system. But, paradoxically for those who want to read philosophical significance into Gödel's theorems, the *proof* of Gödel's incompleteness theorem is, in effect, a proof of the truth of a nonprovable true statement! That is, Gödel's theorem is a normal mathematical theorem

---

How to Prove Everything

Suppose that there exists a statement S such that both S and its negation are provable. We will prove an arbitrary statement R. Notation: $\neg$ denotes negation and $\rightarrow$ denotes (material) implication. We assume the truth of two simple axiom schemes $\neg A \rightarrow (\neg B \rightarrow \neg A)$ and $(\neg B \rightarrow \neg A) \rightarrow (A \rightarrow B)$, for any statements A and B. Here is the proof, using *modus ponens* as the rule of inference from one step to another.

- From the first axiom $\neg S \rightarrow (\neg R \rightarrow \neg S)$ and the assumption that $\neg S$ is provable, we deduce $\neg R \rightarrow \neg S$.

- From the second axiom $(\neg R \rightarrow \neg S) \rightarrow (S \rightarrow R)$ and $\neg R \rightarrow \neg S$ proved in the first step, we deduce $S \rightarrow R$.

- From $S \rightarrow R$ proved in the second step and the assumption that S is provable, we deduce R.

---

that he had to prove to the satisfaction of other mathematicians, otherwise it would never have been acceptable for publication. Gödel simply used proof techniques that are not formalizable within the same deductive system. Similarly, although he proved that the consistency of arithmetic is not provable within the theory of arithmetic, consistency is provable using more advanced mathematical techniques from the theory of infinite sets that are not formalizable within arithmetic.

Surprisingly, all mathematicians living in the 1930s did not thereupon quit their jobs just because mathematics had been shown to be incomplete. Despite its incompleteness, mathematical research thrives and mathematics remains the base of all our science and technology. The reason for the continuing success of mathematics is to be found in the phrase "within the system" that is often ignored in philosophical portrayals of Gödel's work. The theorems are fascinating and surprising, but nothing for a mathematician to write home about, because all they mean is that you have to find some other system in which to prove these statements. So in the end, Gödel's theorems are extremely interesting and a disaster for Hilbert's dreams for the foundations of mathematics, but they hardly caused a ripple in mathematics itself. Certainly, there is no reason to use them to justify a philosophical position that mathematics itself has destroyed science and mathematics.

> **Would it Really Matter?**
>
> Gödel's theorem shows that if a deductive system D is consistent, then D's consistency cannot be proven within D. Suppose, to the contrary, that for some consistent D, its consistency *can* be proven within D. But we have shown in the previous sidebar that if D is *not* consistent, it can prove anything, including its own consistency! So a proof of consistency D within D really proves nothing useful about the consistency of D, because such a proof would exist even if D were inconsistent. Smullyan compares this to trusting a person just because he claims that he never lies. If he is telling the truth, then he never lies and what he says is true, but if he is lying, then he still claims that he never lies, so there is no useful information in the statement.[8]

What is mathematics?

We have not yet asked the obvious question: "What is mathematics?" As with science itself, you can be either a realist or an idealist about mathematics. A realist holds that mathematics, with all its axioms, theorems, and proofs, really exists, and that the task of mathematicians is to *discover* it, just as gravitation exists and it was the task of physicists to discover its laws. Paul Erdős (1913–1996), perhaps the greatest mathematician of the twentieth century, was a mathematical realist. The highest compliment he could pay a colleague was to say that his proof was from "The Book," as if there existed (in heaven?) a book containing all the best proofs of all mathematical theorems. The alternative view holds that mathematics is ultimately a game *invented* by mathematicians. As in chess, where you build a set of artifacts—the game board and the pieces—and then specify a set of rather arbitrary rules by which to play, so it is in mathematics, where you invent concepts and notations and then specify some arbitrary axioms and rules of inference that govern how you play the game, proving theorems.

The game of chess is interesting and challenging, but that does not make the game "real," in the sense that if all knowledge of chess were destroyed, there is no reason to believe that it would be reinvented in precisely the same form. To diagnose whether you are a realist or an idealist about chess or mathematics, ask yourself: If advanced extraterrestrials were to land on Earth, would you expect them to be able to play chess the way we do? If advanced extraterrestrials were to land on Earth, would you expect them to know the mathematical results we know like Gödel's theorems and Fermat's Last Theorem?

While it cannot be denied that the particular notations used in mathematics are invented, that its historical development has been contingent, and that the set of problems considered "interesting" is determined by the society and culture of mathematicians, most mathematicians would subscribe to the realist approach. There are many reasons for this. First and foremost of course, few people (except professional chess players, I suppose) would exert themselves to such an extent for a mere mental game. As everyone knows, mathematics is very difficult, demanding long, frustrating, exhausting hours of solitary toil.[9] Such effort is worthwhile only because we are intensely curious about the world, or because we believe that our work will be applicable.

Furthermore, society supports mathematics in terms of faculty and research positions far beyond the minimum needed to teach elementary math to engineers and accountants. This would hardly be done if mathematics were only a game. I don't recall ever seeing advertisements for professors of chess, and any support that the game gets is similar to that given to sports or culture, not to academic and research activities. One reason for such support is that mathematics, even advanced mathematics, has proved to be useful, because the main theories in science, especially in the physical sciences, are expressed in mathematical laws that have lost any intuitive meaning.

Another reason why most mathematicians are realists about mathematics is that surprises keep turning up in unexpected places. Theories developed for one area of mathematics turn out to be useful in other areas, and seemingly unrelated concepts turn out to be related. (For a simple example, see the sidebar.) It is as if the solution of a chess problem turned out to be essential for a significant advance in games like bridge or poker. You almost have to feel that these connections have been set up beforehand.

Such surprises have also surfaced in the use of modern advanced mathematics to solve classical problems that have been with us for centuries. The ancient Greeks and their successors throughout the middle ages were quite accomplished at the geometrical constructs with a compass and straightedge that you so laboriously performed in high school, and yet there were four problems that they were not able to solve: squaring the circle, trisect-

---

Surprising relations in mathematics

You may have studied the equation $e^{i\pi} = -1$ in a mathematics course in high school or college. It relates four of the most important numbers in mathematics, though they were all developed at different times and for different purposes. Negative numbers like -1 have been used only since the Renaissance, while π was known to the ancient world in equations for the circumference and area of a circle. The mathematics of imaginary numbers like i and the constant e, *base of natural logarithms*, were worked out in the nineteenth century; they are extensively used in mathematical physics.

---

ing an angle, doubling the cube, and constructing a regular heptagon (a polygon with seven sides). Though all these problems can be expressed in elementary high-school geometry, the proofs that these constructions are impossible depend on mathematical discoveries of the nineteenth century. Another fascinating example of the use of modern mathematics to solve a classic problem is discussed in the biographical vignette of Andrew Wiles (1953–) following this chapter.

The presentation of mathematics

Let us return to the issue of linear thinking mentioned at the end of chapter 1. As a general rule, mathematicians aspire to develop the most general concepts and theorems, and the presentation of mathematics in textbooks is almost invariably in a linear sequence from the general to the particular: first some axioms, then a bunch a lemmas and theorems, and finally a set of examples that illustrate the ideas. Students get the idea that you are supposed to solve problems the same way, and then get frustrated when their first attempt at solving a problem fails.

It is a dirty little secret of mathematics that mathematicians work in the opposite direction. They first try out lots of examples to get a feel for the problem, then they may work from the axioms forward, from the problem backward, or even bite off little pieces or simple cases that seem tractable. This is a profoundly complex creative process. But, when it all seems to work, mathematicians cast the solution or proof in the well-structured, linear format of definition-axiom-lemma-theorem-proof-example and carefully shred all evidence of their struggle. The development of mathematics tends to obscure the complex nature of the practice of mathematics. For example, modern algebra grew out of the study of permutations and equations that were then generalized to the study of groups and fields, but now permutations and equations are presented merely as illustrative examples of the general theories. Beginning students often believe that these abstractions were arbitrarily pulled out of a hat by a linear-thinking genius.

Mathematics and science

It is clear today that modern science developed when people stopped debating metaphysical questions about the world and instead concerned themselves with the discovery of laws that were primarily mathematical. We now know the formula for the force of gravity (whatever that is), in terms of mass (whatever that is) and distance (whatever that is). The foundations of modern physics in particular, with its probabilistic quantum mechanics, field theories, and elementary particles, has totally lost touch with our intuitions about the world, and physics has become identical with its mathematical formulation. As Heinrich Hertz said: "Maxwell's theory consists of Maxwell's equations."[10]

Pioneering physicist James Rutherford (1871–1937) has been quoted as saying: "Science consists of physics and stamp collecting." While this is an extreme statement, it has a grain of truth in that the mathematization of physics has become a model for other sciences. Even biology has turned more and more to mathematics. Darwin argued for the theory of evolution by natural selection in qualitative terms, but the modern synthesis is based upon the mathematics of population genetics, and this has been extended into ecological modeling.

In a sense, science is no longer about investigating our direct experience of the world, but more about investigating the mathematical theories that describe it (cf. the biographical vignette of Edward Witten on page 215). If you are at all a realist about the physical world, it would seem that you have no choice but to be a realist—more or less—about mathematics.

* * *

Logic used to be the cornerstone of education, but has somehow disappeared from school curricula. How can you judge if a scientific theory explains and predicts a phenomenon if you cannot carry out the logical inferences required without falling into the traps of logical fallacies? People would be less gullible about according scientific status to creationism if they recognized that creationism is based upon the fallacy of bifurcation.

It is mathematics that enables scientific theories to be used to make precise and accurate predictions about natural phenomena. The choice of Galileo as the first "modern" scientist is not arbirary; it is based upon the recognition of his pioneering use of mathematical models and methods in physics. Scientific research does not have to use mathematics to be interesting or useful, but mathematics does enhance the intellectual depth and practical applicability of a scientific theory.

ANDREW WILES: NEVER SAY UNSOLVABLE

When he was ten years old, Andrew Wiles (1953–) stumbled upon a description of *Fermat's Last Theorem* (sidebar) and determined (like many other budding mathematicians) that he would solve the problem.

Fermat's Last Theorem

In high school, you learned that the equation $x^2 + y^2 = z^2$ has many solutions; for example, $x = 3, y = 4, z = 5$ is a solution, because $3^2 + 4^2 = 9 + 16 = 25 = 5^2$. This is the Pythagorian equation relating the lengths of the sides of right triangles. Fermat considered the generalization of this equation to $x^n + y^n = z^n$ for *arbitrary n*, and conjectured that it has no integer (whole number) solutions for $n > 2$.

Pierre de Fermat was a brilliant seventeenth century mathematician with a penchant for secrecy, and he never wrote out his results or published papers. Instead, he would often summarize his results in marginal notes. Fermat learned his mathematics from the *Arithmetica*, a book of mathematics written by Diophantus (ca. 200–284) of Alexandria. In the margin of his copy of the *Arithmetica*, Fermat wrote out a conjecture and followed it with with the cryptic remark: "I have a truly marvellous demonstration of this proposition which this margin is too narrow to contain."[11]

Fermat wrote many such unproved conjectures, but over the years, all had been proved except for this one, so it came to be known as Fermat's Last Theorem, although it was neither a theorem (because he hadn't proved it), nor his last (just the last claim to remain unproved). The conjecture is exciting because its statement is so simple that any high-school student can understand it, but it had withstood all attempts at proof by the best mathematicians for hundreds of years.

As an adolescent and an undergraduate, Wiles attempted to prove the Theorem, but these attempts were not fruitful. Later he enrolled in graduate studies in mathematics at Cambridge University, specializing in a branch of number theory called elliptic curves. In 1984, Gerhard Frey (1944–) had claimed that if the *Taniyama-Shimura Conjecture* were true, then it would imply the truth of Fermat's Last Theorem. The Conjecture, proposed by Yukata Taniyama (1927–1958) and Goro Shimura (1930–), related two entirely different classes of mathematical entities: elliptic curves and modular forms. Frey's proof was flawed, but two years later Kenneth Ribet (1948–) was able to give a correct proof. Wiles, by now one of the foremost experts on elliptic curves, decided that he would prove the Taniyama-Shimura Conjecture (thus proving Fermat's Last Theorem), despite the belief of most mathematicians that a proof of the conjecture would take decades of research by many mathematicians to complete.

Wiles undertook the proof by himself and in 1993, after seven years of concentrated effort, he announced the proof at a conference in Cambridge. Unfortunately, when the proof was checked by other mathematicians during the peer-review process, a serious error was uncovered. It took another year of intense work and collaboration with Richard Taylor (1962–), before the proof was accepted as correct.

By the time that Wiles had completed his proof, he was just over age forty and thus ineligible to receive the Fields Medal.[12] However, at the awards presentation in 1998, he was granted a special award in recognition of his achievement.

The story of Andrew Wiles's lonely seven-year struggle to solve Fermat's Last Theorem has been documented and is particularly instructive in its descriptions both of the effort needed to make scientific advances and of the social aspect of scientific work.[13] Clearly, Wiles is a brilliant mathematician, but brilliance is not enough; the solution took years of painstaking and frustrating work. Before attempting to work on the problem, Wiles spent eighteen months of preparation, reading mathematical books and journals, and practicing mathematical techniques that would possibly be relevant.[14]

Wiles's decision to work alone was unusual, as most scientific work, even theoretical work, is done collaboratively. The checking and rechecking of the proof by other mathematicians, leading to the discovery to the serious error, demonstrates both the fallible nature of the process of science, as well as its self-correcting mechanisms. The discovery of the mistake in the proof was a difficult psychological blow after being feted for solving a problem where others had failed for hundreds of years, but there was no question that the only course was to attempt to fix the problem.

12 Geology: Modern Science on the Rocks

I used to look down upon geology; after all, what could possibly be interesting about chipping off rock samples with a hammer. While writing this book, I have come to understand that geology is a complex and vibrant science, and that too little of its nature as a science is known to the general public. Since modern geology is of recent origin—much more so than the classical trio of physics, chemistry, and biology—I feel that it is worthwhile devoting a chapter to the nature of geology.

It is often difficult to get an intuitive feeling for the nature of science, because so much of the pioneering work was done hundreds of years ago when people lived quite differently than we do today. Even Einstein's theory of special relativity was published only in 1905, just two years after the Wright brothers' first flight and ten years after the Lumière brothers' first demonstration of cinema. But theories of modern geology are much more recent: *continental drift* was proposed by Alfred Wegener (1880–1930) in 1915, but not accepted until after the theory of *plate tectonics* was proposed by Henry Hess (1906–1969) in 1962. As in other scientific fields, we shall see the same interplay of theory and observation, of initial rejection and ultimate acceptance that we saw in other earlier theories, and we can interpret this process within our definition of a scientific theory.

Perhaps since the origin of the Earth has less personal implications for our self-esteem than the origin of life, theories in geology have evoked somewhat less conflict with religion than have theories of evolution. Many creationists are willing to forego the literal truth of creation in six days, but continue to believe in the special creation of each species. Nevertheless, the study of geology is extremely similar to the study of evolution because they are both primarily historical sciences. Even though the processes described by the theories continue to be active today, the rate of change of observable phenomena is so slow that it takes some effort to assimilate their concepts.

Furthermore, both geological theories and the theory of evolution are used primarily for retrodiction. The primary aim of the theory of evolution is to explain the origin of the vast number of species of plants and animals—both living and extinct—and the relationships among the various species, while the primary aim in geology is to explain the origin of the features of the Earth that we see—mountains, rocks, and oceans—and to investigate the composition of the interior of the Earth.

Both fields are also relevant to the future: the theory of evolution is the basis of concern about biodiversity and the development of drug-resistant microorganisms, while theories of geology can explain potential natural catastrophes like earthquakes and volcanic eruptions. During the past few decades, geologists have applied their knowledge in the exploration of space because the principles that explain the structure of the Earth are used to explain the structure of planets and their moons. Evolutionary biology has yet to find application outside the Earth, but you can never know.

The study of geology

The tools of evolutionary biology come from the study of anatomy and physiology. From a knowledge of the anatomy and physiology of modern-day reptiles and birds obtained from field work and laboratory experimentation, a scientist can deduce the anatomy and physiology of extinct dinosaurs, even though he must work from fragmentary fossilized bones. Fortunately, the basic principles of life have not changed throughout hundreds of millions of years, so the reconstructions are quite reliable. Sometimes scientists disagree, for example, on the question of whether dinosaurs were warm-blooded or cold-blooded, but the disagreement is a matter of interpretation of the meager clues that have been left in the fossil record and could be resolved if sufficient evidence becomes available.

The study of the development of the Earth is much more difficult. We don't have another Earth-like planet that can be conveniently studied in the laboratory in order to obtain results that could be used to reconstruct the origin and composition of the Earth. Geology, more so than biology, depends on the interpretation of basic observations, but since the Earth is so large, these will necessarily be sporadic. The surface area of the Earth is about 500 million square kilometers, two-thirds of which are covered by oceans, so even a couple thousand samples drilled from the Earth barely scratch the surface. Furthermore, even basic observations are hard to obtain, because the behavior of rock under the immense pressures and temperatures that exist within the Earth is not something that can be easily duplicated in the laboratory. Finally, the really interesting places that you would want to investigate—active volcanoes and the deepest trenches in the ocean floor—are not exactly vacation paradises that are easily accessible.

Geologists must work with clues obtained from four sources: investigations of the chemical and physical properties of rocks, surveys of the structure of features of the Earth's surface, indirect evidence of the composition of the interior of the earth obtained by instruments such as seismographs, and contributions from allied sciences like meteorology and oceanography. There is a fruitful positive-feedback loop between geology and evolutionary biology. If you find significant deposits of marine organisms, you can conclude that the rocks were created in the sea; similarly, if you find cold-blooded reptiles, you can be sure that the rocks were not created at the poles. In the other direction, the temporal relationships among fossils can be established by studying the geological structures within which they are found, as well as by radioisotope dating of the fossils. Once you establish a date for certain fossils, you can assign a date to any geological structure anywhere on Earth where such fossils are found. In particular, small organisms are extremely common and they change their structure quite frequently, so biological dating can be quite precise.

As early as the eighteenth century, Pierre Simon de Laplace (1749–1827) had proposed that the Sun and the planets coalesced from rotating gas and dust, gradually becoming hotter as gravitation contracted the spinning material. The result would be a more or less smooth, more or less spherical, Earth.[1] So where did all the oceans and mountains come from? The number of possible explanations is large. If the Earth cooled to form a crust and then contracted again, the crust would become wrinkled like a dried apple. If the Earth cooled, formed a crust, and then expanded, cracks would open up like they do in a less-than-perfectly-baked sponge cake. Since different types of rock have different densities, blocks of land could sink or be pushed up, especially if the interior of the Earth is liquid.

It is easy to suggest a hypothesis, but it is extremely difficult to construct a concise, coherent theory that can be used to explain and predict. As in other sciences, a circularity exists: geologists need data on the composition of the Earth in order to develop theories of its structure, and conversely, they need theories to guide observations. The composition of the interior of the Earth can only be inferred from clues obtained from the analysis of seismic data from Earthquakes and explosions, and high-quality data of this sort has become available only since the second half of the twentieth century. In particular, theories that are based on uplifting or sinking or movement of continents depend on knowledge of the structure and composition of the ocean floor, but until very recently the technology did not exist for obtaining such data. Modern geology depends not only on brute-force techniques for drilling and sampling, but also on electronics and computers that enable scientists to collect, collate, and analyze large amounts of data. Computers are also essential as surrogate geological laboratories. Since you can't run experiments to investigate predictions of alternate theories for the origins of features of the Earth, the only option is to perform mathematical modeling of alternative theories with the aid of a computer.

Wegener and the theory of continental drift

In retrospect, the decisive moment for modern geology was the publication of *The Origin of Continents and Oceans* by Alfred Wegener in 1915. Wegener proposed that all the continents were originally parts of a single landmass that broke apart; then the continents drifted until they arrived at their present locations, and in fact the drift continues to this day. Every since the "discovery" and mapping of the "new" world, people had noticed that the coastlines of the Americas could be matched with the coastlines of Europe and Africa. But Wegener was the first to marshal evidence to support a concise and coherent theory of continental drift.

The evidence comes from three sources. First, not only do the coastal outlines of the continents match, but the detailed geological formations on opposite sides of the Atlantic ocean match as well. Geologists from South Africa feel right at home in parts of Brazil, and the Appalachian mountains in the United States are geologically very similar to the mountains of Scotland and Norway. Second, an analysis of both living and extinct animals and plants shows many cases in which small pockets of similarities are found at extremely remote localities, but not in neighboring localities. Why are there lemurs in Madagascar and related *prosimian* species in locations across the Indian ocean? Why are there marsupials in Australia and America, but not in New Guinea or South East Asia (a relative stone's throw from Australia)? Findings of rare extinct animals provide even better evidence for ancient connections between continents. A remarkably ugly reptile called *Kannemeyeria* lived in both Africa and Argentina, but it was simply too large and lumbering to have crossed the Atlantic and as a cold-blooded reptile could not have walked over the frigid land bridge between Siberia and Alaska on its way from one hot climate to another.

Third, there are fossil finds that are totally inconsistent with the current geography of the Earth. The Himalayan mountains are far from any ocean, but they are primarily built of sedimentary rock filled with the skeletons of marine animals. Therefore, the mountains must have been formed of land that was originally part of the sea floor. Similarly, coal deposits and fossils found in Antarctica indicate that the continent once had a lush climate, though it is currently capable of supporting only organisms that are specially adapted for its harsh, cold climate.

Analyzing all these data, Alfred Wegener concluded that the continents had once been joined and had drifted apart. The currently accepted theory proposes that a single landmass called *Pangea* broke apart about 200 million years ago into two supercontinents: *Laurasia* comprising (present-day) Asia, Europe, and North America, and *Gondwanaland* comprising (present-day) Africa, South America, India, Australia, and Antarctica. These then split into the continents we recognize today and "drifted" apart to their current positions. In fact, the continents are still moving at the rate of a few centimeters per year, as confirmed by extremely precise measurements made with the aid of laser reflectors placed on satellites and the Moon.

The theory of continental drift has been able to explain many of the puzzling observations mentioned above. The Himalayas were formed when India was detached from Gondwanaland and moved rapidly north to crash into Asia, pushing up ocean sediments into a vast mountain range in the same way that the hood of an automobile gets crumpled in a collision. Antarctica was originally located far north of its present position when it was part of Gondwanaland for millions of years, supporting tropical plants and animals that were eventually squeezed into coal. Only about thirty million years ago did it drift to the South Pole and begin to develop its massive ice cap.

They laughed at Alfred Wegener. Well, not really and not everyone. But the majority of professional geologists regarded his theory as absurd.

Part of the reason was that he was an outsider, and part of the reason was that other geologists had their own theories that explained and predicted the same phenomena. The change in the climate of the continents reflected by the changes in the fossil record could be explained by assuming that the Earth as a whole had changed its orientation in space, that is, that the geographic North and South Poles (defined by the axis of rotation of the Earth) had moved. It would be years before internal inconsistencies and theoretical calculations were able to refute this possibility. The existence of related species far removed from each other could be explained by assuming that there were land bridges between the continents that had subsequently sunk.

But the main reason for the cool reception accorded continental drift was the lack of a plausible mechanism. The best Wegener could do was to assume that continents "plowed" their way through the oceanic crust, but that made no sense because it was known that the floor of the oceans was composed of hard rock. For several decades after its proposal, the theory of continental drift languished, both because it lacked both an explanatory mechanism and also because a large body of its predictions had alternative explanations. Not that the other theories were much more successful, but at least they did not have the disadvantage of being counterintuitive. It is easier to accept a contracting or expanding Earth than to accept that continents plow through hard rock.

Vindication by plate tectonics

Following World War II, there was a massive increase in the funding for undersea exploration as well as significant improvement in the available technology. This can be attributed to commercial and military interests. Petroleum companies wished to extend their search for offshore oil fields and navies wished to improve their ability to maneuver their own submarines and to detect enemy ones. By the 1960s, the accumulated evidence led geologist Henry Hess to propose a new model for the Earth called *plate tectonics*. According to this model, all continents originally formed a single landmass, but large cracks in the Earth's crust allowed hot lava to well up from within. The lava cooled and formed plates, which pushed apart other plates, including those upon which the continents rode. Since the continents are formed of lighter rock, eventually the ocean plates sank underneath the continents to melt again. A large upwelling coming from the Mid-Atlantic Ridge has pushed the American continents away from the plates upon which Europe and Africa ride. Continents do not drift as Wegener supposed; rather, they are simply being pushed apart. However, Wegener's reconstruction of the movements of the continents is supported by the structure of the oceanic and the continental plates.

The evidence supporting plate tectonics is massive. All rock samples taken from the ocean floors are no older than 200–300 million years, while on the continents, rocks can be found that date from close to the origin of the Earth billions of years ago. Samples from mid-ocean ridges are relatively young, while samples taken near the deep trenches that are found close to the continents are relatively old. There is relatively little sediment near the mid-ocean ridges, while relatively deep deposits are found near the continents. The theory of plate tectonics offers a simple explanation of these data. The ocean floors consist entirely of rock that has hardened as it welled up and was then transported with its plate. Younger rock is lava that has hardened recently, while older rock has moved toward the trenches. Rocks older than about 200 million years have simply vanished into the Earth's interior as the ocean plates slide under the continental plates. There has been little time for sediment to accumulate on the younger rocks near the mid-ocean ridges, while the older rocks have been exposed to sedimentation for much longer periods.

Still, other theories can explain these observations. Sediments can be swept away by currents, or the rate of sedimentation can change with the properties of the water of the oceans. Eventually, the mass of observations became inconsistent with the other theories, while remaining consistent with plate tectonics, but before that, a new set of observations appeared that proved to be decisive in favor of plate tectonics. When new rock is formed by the hardening of molten lava, magnetic material aligns itself with the Earth's magnetic poles. For reasons that are not fully understood, the direction of the Earth's magnetic field reverses itself frequently (frequently, that is, in geological terms, every few million years). Once a rock hardens, the direction of its magnetism can no longer be changed, so such rocks form what can be called fossils of magnetism.

Paleomagnetism is the study of the ancient magnetic properties of the Earth, obtained by computations and measurements performed on samples of rocks containing magnetic material. In 1963, Frederick Vine (1939–) and Drummond Matthews (1931–1997) noted that plots of the magnetic properties of the ocean floor show an amazing property. The patterns of magnetic reversal on either side of mid-oceanic rifts are mirror images of each other. The obvious explanation is that the rocks were formed at the site of the rift and then rode opposing plates as they are pushed apart. Initially, some geologists did not accept the results from paleomagnetism as evidence for plate tectonics, because the diagrams look almost like inkblot tests used by psychologists, and it was suspected that the alleged symmetries were the product of wishful thinking. In time, careful presentation of the accumulated evidence won over almost all geologists.

Mechanism, or explain and predict

The fate of Wegener's theory of continental drift can be compared with the fate of Newton's theory of gravitation. In both cases, the theory was proposed without a mechanism, and in both cases there were flaws in the theory that had to be corrected later by plate tectonics and general relativity, respectively. Yet Alfred Wegener was laughed at, while Isaac Newton was lionized. Why were their fates so dissimilar? There are two reasons. First, gravitation is familiar and it is especially easy to observe and measure phenomena that can be explained by gravitation, such as planetary motion, pendulums, and projectiles. Second, gravitation is easily amenable to a mathematical treatment. Newton and his successors were able to give explanations and predictions that were so amazingly accurate and precise that no one seriously considered questioning the basic theory for over two hundred years. Wegener certainly marshaled evidence for continental drift, but because of the nature of the science of geology it was necessarily fragmentary and equivocal. Only after plate tectonics was proposed as a mechanism for continental drift did Wegener's theory garner support, and eventually all or almost all serious critics were won over by the preponderance of evidence in its favor.

This complete turnabout during a decade or so from the time that Hess proposed tectonics is a counterexample to Kuhn's claim that scientists do not change their minds based on new evidence, but that old theories die out only when their proponents die out. Paradoxically, Wegener's theory was not vindicated until a viable mechanism was available, but the mechanism itself, the theory of plate tectonics, has been accepted despite the fact that there is no agreement on *its* mechanism! The immense forces required to move the ocean plates almost certainly come from currents of molten rock within the Earth, but no fully satisfactory explanation exists yet. Several possibilities do exist: plates are pushed apart by the upwelling of lava in rifts; plates are pulled down by the weight of the cool, hardened rock sinking into troughs; plates are dragged along by the convection currents in the mantle; massive amounts of molten material welling up from relatively small *plumes* (100 kilometers in diameter) cause the motion of the plates.

Continental drift had to wait for a mechanism to be worked out, but plate tectonics, like the theory of gravitation, is so successful in explaining and predicting that it is accepted despite the lack of agreement on the precise mechanism that causes it.

* * *

Historical sciences like cosmology, geology, and evolutionary biology do not fit the naive view of scientists proposing scientific theories and then carrying out experiments to confirm or falsify them. Experiments are impossible and empirical data is hard to obtain and fragmentary. However, this does not mean that these fields are not scientific, and that their theories do not need to conform to the definition of scientific theories. It does mean that predictions become retrodictions and that a long time may pass between the proposal of a theory and the availability of data to check its retrodictions. In fact, such data may never be forthcoming. For these reasons, the ability to provide a mechanism for a theory is important, not only because it makes the theory more plausible, but also because mechanisms are at a lower level of abstraction (basic physics, chemistry, or biology) and thus more firmly established. Rejecting Wegener's theory of continental drift was not unreasonable until the mechanism of plate tectonics was suggested and supported by experimental results.

ALFRED WEGENER: STEADFAST IN SCIENCE AND ON THE ICE

If there were ever a pioneering scientist who was the total opposite of the stereotyped nerd starved for exercise and fresh air, it must have been Alfred Wegener (1880–1930). Wegener obtained his doctorate in astronomy in 1905, and a year later made a recording-breaking fifty-two-hour balloon flight. His interest shifted to meteorology and in 1913 he crossed the Greenland ice cap. (This was only two years after Roald Amundsen's [1872–1928] expedition first reached the South Pole.) In 1914, Wegener served as an infantry lieutenant in the German army and was wounded twice. The second wound was serious enough that it demanded a long period of convalescence and his release from further service. It was during this period that Wegener deepened his research into continental drift and published the first edition of *The Origin of Continents and Oceans*.

Wegener's interdisciplinary interests and research made it quite impossible for him to obtain an academic position in an age where the concept of "interdisciplinary" did not exist. It was not until 1928 that he was finally offered a position especially created for him at the University of Graz in Austria.

In 1930 he led another expedition to Greenland that was to establish a station in the middle of the ice cap to perform meteorological and geophysical observations during the winter. Two of his colleagues were already at the station, desperately in need of supplies before they were cut off by the winter weather. Despite serious delays and worsening weather, Wegener refused to abandon his friends, eventually arriving at the station with two companions, one of whom was so severely frostbitten that he had to remain. The supplies would not suffice for everyone, so the next morning, Wegener set out on the return journey with Rasmus Willumsen, a native of Greenland. Wegener apparently died of a heart attack and his body was recovered the next summer; Willumsen was never seen again.

One of the claims frequently made by opponents of science is that scientists organize themselves into closed guilds, and that outsiders with revolutionary ideas are not afforded a cordial reception. This is both true and false in the case of Wegener. It is true that since he was not a geologist (he was trained in astronomy and later made the transition to meteorology), professional geologists considered him as an interloper writing in a field that was not his. Nevertheless, Wegener was a bona fide scientist, and he knew how to present his ideas to scientists by marshaling evidence, drawing conclusions, proposing tests, and responding to criticisms.

Like Darwin, Wegener searched for evidence wherever he could find it: geological structures on the Earth's surface, geophysical theories of the composition and movements of the interior of the Earth, studies of the distribution of flora and fauna throughout the world, and the implications of the fossil record on climate change. Also like Darwin, Wegener revised his *Origin* several times, modifying his theory in response to criticism and new evidence, and would certainly have prepared further revisions had he lived longer. Just as Darwin courageously confronted possible difficulties with the theory of evolution, Wegener was deeply conscious of the problems with his theory of continental drift, above all because of the lack of a reasonable mechanism to explain the drift. "The Newton of drift theory has not yet appeared," he wrote.[2]

Had Wegener lived to a ripe old age, he would have seen the emergence of plate tectonics as a mechanism and the beginning of the triumph of his theory.

13

The Future of Science: Surprises or Revolutions

The end of science?

The phenomenal success of science in the twentieth century has given rise to two forecasts as to the future of science. Some see science continuing to progress, perhaps even at an ever-increasing rate, while others look back on the achievements of the previous century and judge that the great advances have already been made. This debate has been popularized by two books: *What Remains to Be Discovered* by John Maddox (1925–), the former editor of *Nature*, one of the preeminent scientific journals, and *The End of Science* by John Horgan (1953–), a writer for *Scientific American*. The future of science can also be framed within the philosophy of science expressed by Thomas Kuhn (chapter 6): Are scientific revolutions inevitable? Is another scientific revolution likely? This final chapter contains my thoughts on the debate.

The seminal anecdote in the debate on the future of science occurred over a century ago in 1894. American physicist Albert Michelson (1852–1931) gave a speech in which he said: "An eminent physicist has remarked that the future truths of Physical Science are to be looked for in the sixth place of decimals."[1] At the end of the nineteenth century, *some* scientists believed that they had solved most of the important problems and that future scientists would only be able to provide minor corrections to existing theories. Ironically, in 1887 Michelson himself, together with Edward Morley (1838–1923), had already performed a experiment showing that the speed of light was constant and not affected by the movement of the Earth. This experiment showed that there were unexplained phenomena in physics that went beyond the "sixth place of decimals." Just eleven years after this speech, Albert Einstein published his papers on special relativity and the quantum nature of light, precipitating a revolution in physics.

Parenthetically, it is important to note that Michelson was wrong, in the sense that not all scientific questions were solved to the satisfaction of scientists of that day. For example, there was the glaring difficulty that stood in the way of the full acceptance of Darwin's theory of evolution by natural selection. The best estimate of the age of the Earth had been given by William Thompson (Lord Kelvin) (1824–1907), who calculated the time it would take for the Earth to cool down to its present temperature. The value he calculated was far too short to allow life as we know it to evolve, and yet even then, the theory of evolution was strongly supported by the evidence. It would not have been out of place to expect that a revolutionary discovery would resolve the clash between the theories of evolution and thermodynamics. In fact, a few years later such a revolutionary discovery did occur. The discovery of radioactivity enabled scientists to show that heat produced by radioactive material within the Earth was sufficient to explain how the Earth could be old enough for evolution to take place.

For Maddox, historical precedent is sufficiently convincing to justify a belief in future revolutions in science: "History is on the side of the second camp [those who believe that a 'new physics' will be thrown up instead of a 'theory of everything'], to which I belong."[2] Nevertheless, history does not necessarily repeat itself; for example, the Byzantine Empire lasted (nominally) for over one thousand years from its founding by Constantine in 324 to its final destruction when Constantinople was captured by the Ottoman Turks in 1453, but attempts by Napoleon and Hitler to establish long-lasting empires quickly failed. So it is appropriate when speculating on the future of science to consider the possibility that history may not repeat itself.

John Horgan, in *The End of Science*, comes down forcefully on the side of those who do not foresee any revolutions in science that are comparable to those that took place in the twentieth century. Science, in the form of concise and coherent theories that can be used to explain and predict, has become the victim of its own success. There is a consensus that the theories of physics are true down to the level of quarks and back to the first second after the big bang. Chemistry is firmly based on quantum mechanics, and the biochemistry of molecular biology furnishes a mechanism for explaining the processes of life. Horgan interviewed many brilliant scientists who have proposed radical new approaches to science, but they are all speculation at this point, not even close to being concise and coherent, and even further from being able to explain and predict natural phenomena. Notwithstanding the embarrassment that Michelson's speech holds for prognosticators, it is possible that history will not repeat itself and that there will be few, if any, *revolutionary* scientific discoveries.

What would a new scientific revolution look like?

Even if some of the Big Questions of science are solved, the reaction might well be "So what?" Maxwell's theory of electromagnetism led to an electronics industry that is bound up intimately with every aspect of our lives. Quantum mechanics, as applied in improving our understanding of chemistry and our ability to design semiconductor devices has surely influenced our lives, usually for the better, while nuclear physics poses both a threat to our existence and a promise for a better life if controlled fusion is ever able to provide us with clean, inexpensive energy. But who is really affected by modern quantum field theories? So what if nucleons are composed of quarks exchanging gluons? The predictive power of the theory is limited to phenomena that can only be observed in extremely large and expensive experiments.

Suppose now that a concise and coherent theory were developed that was able to explain the ultimate composition of mass and energy, time and space. A current candidate is the theory of *superstrings*, which proposes that everything is composed of ten-dimensional packets of vibration called strings though we can sense only four of the dimensions.[3] The image on the next page is a rendering of a six-dimensional Calabi-Yau manifold.[4] You have to try to imagine that each "point" of four-dimensional space-time *is* this thing. Be careful not to say that each point "contains" a Calabi-Yau manifold, because space and time exist only as manifestations of their properties. Strings are inconceivably small. The unit of measure of strings is the *Planck length*, which is about 10^{-35} meters. For comparison, the radius of an electron, the smallest particle that is relevant to everyday technology, is about 10^{-15} meters. We find it difficult to imagine how small atoms and electrons are, yet the objects of string theory are even smaller relative to electrons then electrons are relative to us. What chance do we stand of truly understanding superstring theory other than as a mathematical formalism?

What would happen if eventually the preponderance of evidence were to support superstring theory? The theory would explain and predict phenomena that can only be observed in a particle accelerator so large as to be beyond our ability to build it, or very remote phenomena observed in large telescopes. Similarly, a future consensus on the events that happened during the first second of the big bang is hardly likely to affect anyone but a handful of physicists.

In biology, the revolution brought about by molecular biology has been so complete that it is doubtful that there is room for more revolutions. To be sure, a lot is unknown: how an organism develops from an embryo and how the brain works. But the answers, if they are found, will almost certainly be found within the framework of molecular biology as it is currently practiced. Similarly, we may or may not be able to come up with a concise and coherent theory that accounts for the origin of life, the origin of self-replication in DNA and RNA. Maybe God created it ex nihilo. But if we do find a natural explanation, it will probably be a nonrevolutionary biochemical process.

Horgan and Maddox do not entirely disagree; the apparent contradiction is mostly a matter of semantics, of distinguishing between *surprises* and *revolutions*. For example, Maddox states: "Any belief that there are no surprises left should be dispelled by [the discovery of quasars]."[5] It is true that this discovery was a surprise, but it fails miserably as an example of a revolution. Maddox himself describes the theory that quasars are caused by black holes, a concept dating back to 1916 when Karl Schwarzschild (1873–1916) was working out some consequences of Einstein's general theory of relativity. If, eventually, the preponderance of evidence convinces scientists that quasars are in fact powered by black holes, no revolutionary change will occur in the foundations of physics; on the contrary, it will be taken as a confirmation of Einstein's theory that is almost a century old.

Here is an analogy. If a mother bought her son a large motorcycle for his eighteenth birthday that would be a surprise, because mothers usually try to convince teenagers not to drive dangerous vehicles like motorcycles; but it would not be a revolution. The boy would have been asking for the motorcycle for months, and hoping against all reason that he would be granted his wish. Suppose, however, that the same mother told her son on his eighteenth birthday that he had been adopted or born to a surrogate mother. This would be a *big* surprise, so much so that it might cause the boy psychological distress, and for that reason psychologists recommend discussing such matters at an early age. But it still could not be called a revolution, because the boy would have known about adoption or surrogacy from his friends or from the media. If, however, he were told that his origin was from the planet Krypton, now that would be a revolution!

There are other reasons to believe that science is a victim of its own success. First, modern science has progressed by successive refinement, and second, science has offered explanations for almost all familiar natural phenomena. Let us examine these in detail.

Refinement of scientific theories

It is frequently said that old theories are *approximations* of the newer theories that replace them, for example, that Newtonian mechanics is an approximation to Einstein's theory of special relativity. This terminology is very misleading. The theory of special relativity claims that Newtonian mechanics is *not* correct, because relativity denies the existence of the basic concepts of Newtonian mechanics: absolute space and time, unbounded velocity. All that can be said is that Newtonian mechanics is *predictively equivalent* or *computationally equivalent* to special relativity under a large range of conditions.

Suppose now, that according to a new and revolutionary theory, Einstein's theory of relativity is shown to be totally incorrect. Both common sense and historical precedent (many precedents, since the time of Galileo) show that when there exists a massive amount of evidence that supports a theory, then any new theory will be predictively and computationally equivalent to the old theory that it replaces under a large range of conditions. Thus a theory that purports to replace relativity would almost certainly be compatible with relativity, except under certain extreme conditions, perhaps close to the beginning of the big bang or in the internal structure of elementary particles. *Star Trek* fans are truly advised to enjoy the TV shows and movies as entertaining science fiction, and not to wait with bated breath for warp-speed travel of spaceships and transporters of macroscopic creatures.

In biology too, we see that replacement theories are predictively equivalent. Darwin's fumbling efforts to explain heredity were overthrown by Mendelian genetics, but this theory simply sharpened Darwin's theory of evolution by natural selection by giving it a firm and mathematical underpinning. Subsequently, molecular genetics was able to show that the simplistic assumptions of Mendelian genetics did not hold, but the molecular theory was in general predictively equivalent to the Mendelian theory.

Science explains natural phenomena

A second reason for skepticism about an unbounded sequence of scientific revolutions is that most science concerns natural phenomena whose existence is clear and totally uncontroversial. Newtonian mechanics explained the motions of planets and comets that had been observed for eons in all cultures. Solar eclipses occurred frequently and all science did was to replace myths about dragons eating the Sun with theories of motion and gravitation that explain why eclipses occurred and that enable physicists to calculate their occurrence with great precision and accuracy. Similarly, electricity was known to everyone through lightning, and magnetic compasses have been used for ages. If someone had claimed a thousand years ago that lightning would one day be used for practically instantaneous communication, he might have been burned at the stake as a heretic, but—in retrospect, of course—the claim could not be considered outrageous.

Similarly, in biology, the structural kinships among animals and plants were recognized, although an explanation would have to wait for the theory of evolution. Disease was another omnipresent natural phenomenon that one could hope to explain scientifically.

When we come to twentieth-century science, the situation is somewhat different, but it is still possible to claim that science has been explaining natural phenomena whose existence is uncontroversial. We are all intimately familiar with the intense radiation emitted by the Sun, so a physicist in 1900 would not have been too surprised if we had returned in a time machine and told him that physics would discover a new type of reaction that fuels the Sun.

So, are there any natural phenomena that await explanation? There are precious few in our everyday life, especially if we exclude historical questions like: "How did the universe begin?" and "How did life arise?" and "How did the human cognitive abilities evolve?" These questions are of overwhelming interest, but they may be unsolvable simply because there is no longer enough evidence remaining. At most we might be able to offer convincing arguments and experimental results that establish their plausibility, but conclusive evidence may never become available.

There are fundamental questions about natural phenomena that await explanation, for example: "What is consciousness?" and "What causes the movement of tectonic plates?" and "How does an embryo composed of a few cells become a complex organism?" But one can't help feeling that even if science is eventually able to solve these problems, the solutions will not come from revolutionary changes like overthrowing and replacing quantum mechanics or molecular biology, but by applications and extensions of these existing basic sciences.

One can imagine events that everyone would agree to classify as revolutionary. An example would be the discovery of a life form (either here on Earth or from elsewhere) that was not based on the familiar biochemistry of nucleic acids and proteins. Even if an alternate biochemistry exists, it is hard to imagine that it would not be explainable in terms of our chemistry and physics. On the other hand, if an extraterrestrial life form were found that was based upon our familiar biochemistry, the discovery would be astounding, but not revolutionary science.

Expectations from science

When I was young I considered becoming a physicist in order to participate in the most exciting scientific revolution that was being forecast—the achievement of controlled fusion, which promised safe, unlimited, low-cost energy. The breakthrough was just around the corner. Several decades later, the corner is still there and our power plants are still burning oil and coal, and splitting uranium.

The close connection between scientific progress and progress in the application of science has led to an inflation of expectations from science, and to an excess of marketing hype on the part of scientists. It is true that basic science leads to applied science (and conversely!), and it is true that you cannot predict which research directions will lead to improvements in applications. Nevertheless, it is irresponsible of scientists to justify every research project on the grounds that it might help find a cure for cancer. First, it is not true, and second, it leads to unfulfilled expectations and disillusionment. There is no way to predict what research, *if any*, will lead to a "cure" for cancer. After a few decades, if science can only offer measured progress in treatment, rather than a "cure," the tax-paying public will start debating, with a certain amount of justification, the possibility of diverting investment in science to other fields.

The hype over the *Human Genome Project* is an example of such excessive marketing. The project was claimed to have immense implications for medical science. Now that the project is completed, we hear, of course, that research on applications in medicine is only just beginning, and that new treatments may be decades away.

A similar situation is now occurring with *nanotechnology*, which is concerned with creating machines the size of biological molecules. This is about a thousand times smaller than the technology used to pack a million electronic components into a finger-nail-sized "chip." This cutting-edge scientific research in physics and chemistry that advances step-by-little-step is being hyped as an imminent revolution.[6]

People engage in science because it is a human trait to be curious. Like Alexander the Great who kept wanting to see what was beyond the next river until his soldiers refused to to take another step, scientists always want to know a bit more. Coincidently, science has made enormous changes in our lives, mostly for the better, but that should not tempt scientists to excessive and unjustified marketing.

The limits of technological progress

The explosive progress in some fields of technology, particularly in computers, communications, and medicine, should not blind us to the fact that a revolutionary expansion cannot go on forever. General Motors, before closing its Oldsmobile division, tried to convince customers that "this is not your father's Oldsmobile," but the plain truth is that it was. The scientific and technological principles of motor transport go back more than a hundred years since the first engines were built by Nicholas Otto (1832–1891) in 1876 and Rudolf Diesel (1858–1913) in 1897. Cars have not changed much in decades, because the laws of thermodynamics—discovered in the mid-nineteenth century by Sadi Carnot, William Thompson, Hermann von Helmholtz (1821–1894), James Joule (1818–1889), and Rudolf Clausius (1822–1888)—simply refuse to go away. There are continual, often significant, improvements like fuel-injection, electronic ignition, and antilock brakes, but no revolutions. "Futuristic" electric vehicles have been with us for decades in the guise of golf carts and wheelchairs.

Similarly, there is no revolutionary difference between the first Boeing 707 aircraft manufactured in 1954 and a shiny new 747 coming off the production line today. Glass cockpits and winglets (those turned-up surfaces that give the wings an insolent look) do not a revolution make. The relatively revolutionary attempt to build a supersonic aircraft, the Concorde, foundered after a handful of aircraft were built. The laws of aerodynamics and thermodynamics made them playthings of the rich, rather than essential tools of transportation like the mundane machines manufactured by Boeing and Airbus. British Airways and Air France have recently taken the Concorde out of service and there are no plans to develop a replacement. The relevant technologies have stagnated for decades and offer no solutions to the economic and environmental problems that plagued supersonic passenger aircraft.

The history of manned space exploration shows that the progress of science and technology is not necessarily linear or even inevitable. During the Apollo program, men landed on the Moon a total of only six times during the years 1969–1972. Innumerable books and articles were published predicting a brave new world of space travel. Thirty years later, not only has manned exploration of the Moon and planets not progressed, but to my knowledge there are not even any serious plans to resume them during the coming decades.

Let me make a simple comparison. At the time of the Apollo program, I held jobs working on large, expensive mainframe computers, whose computing power was puny compared even with the computer in my office upon which I am typing this manuscript. Why was progress in computing technology so fast compared with the lack of progress in space travel? The reason is very simple: computing technology is only now approaching scientific limits such as quantum uncertainty and the speed of light, while space technology has already run into its limits that derive from the basic principles of physics and chemistry.

Alternative technologies, like nuclear and solar propulsion, have not lived up to their initial promise, and even they are likely to bring only relatively minor extensions of existing capabilities. Incremental engineering improvements in space technology mean that we could return to the Moon perhaps somewhat more cheaply than before, and perhaps travel to Mars, but routine space travel within the solar system, or any travel whatsoever outside it, is likely to remain impossible indefinitely. The same lesson holds for computing technology. The explosive progress in the recent past does not guarantee that it will be maintained indefinitely.

Extraterrestrial life

These considerations also explain why scientists are not terribly interested in the possibility of extraterrestrial (ET) life. The real reason that scientists are so reluctant to become involved in discussions of ET life is not the lack of evidence but the total lack of plausibility and relevance. It is not at all implausible that ET life exists, given that there are at least a million galaxies, each with billions of stars in them. But it is totally implausible that ET life could visit us or that we could visit it, so it doesn't really matter if it exists or not.

The Saturn V rockets propelled the cramped Apollo capsule past the escape velocity from Earth orbit, about 11 kilometers/second. Suppose that an ET life form has an inconceivable, incredible supertechnology enabling them to achieve speeds of 300 kilometers/second, almost 30 times faster than we can. Since the speed of light is 300,000 kilometers/second, to find the travel time between another star and the Earth at speeds of 300 kilometers/second, we have to multiple distances in light-years by 1,000. To get to any star more than 200 light-years away would take longer than the entire existence of our species *Homo sapiens*, and to get to any star more than 4 light-years away (that is, to any star) would take almost as long as the entire length of recorded history. Galaxies are millions of light-years away, so intergalactic travel is totally meaningless. We see that even an unimaginable thirtyfold improvement in space technology would not make space travel practical.

Is it not possible that ET life forms are in possession of a technology that we cannot imagine, a technology that would enable them to infiltrate us over intergalactic space without our perceiving them? The answer is almost certainly no, for the simple reason that the laws of nature are the same everywhere in the universe. How can we possibly know what the laws of physics are in remote galaxies? One of the most essential tools in astronomy is the measurement of the spectra from space. Spectra are formed because the frequencies of light emitted or absorbed by atomic elements are determined by the transitions of electrons between energy levels in atoms. The spectra of all light observed from all stars and galaxies correspond precisely to those of the known atomic elements, and can be explained by the laws of quantum mechanics that have been studied for decades. So we know that the laws of physics and chemistry are truly universal. ET life may exist and it may be based on our biochemistry or on some other biochemistry, but it cannot escape the "laws of nature." The paraphernalia of science fiction—worm holes, time travel, warp drives—may exist on a quantum level, but they do not exist on a macroscopic level, neither for us, nor for any ET life forms.

Furthermore, if ET life exists, it would have to work with the known forces of gravitation and electromagnetism, and with the known atomic elements. It is therefore inconceivable that a highly developed technological ET society exists without a full grasp of electromagnetism. There is little doubt that we will be able to receive electromagnetic emissions from an

ET society before they swoop down to abduct us. If that really were their intention, they would have to be able to identify our planet as a source of potential delicacies, and this would only be possible by locating and analyzing our artificial electromagnetic emissions that began only about a century ago. They have barely had time to move more than a few light years, so eons will pass until they reach us.

Finally, if ET lifeforms really are so advanced compared to us, what possible reason would they have for undertaking a journey of thousands or millions of years just to come here? Few of us would be willing to devote our lives and those of our descendants to searching out some obscure species of insects in a jungle, whether out of scientific curiosity or as a culinary tidbits. It is the height of egocentric thinking to even imagine that an advanced civilization would find us worth investigating!

If we found out that ET life exists, it would be surprising but not revolutionary. If ET life does exist we will almost certainly know about it from electromagnetic emissions long before we actually meet them. Carl Sagan's book *Contact*, in which he describes the reception of signals from space, is thus a very reasonable scenario for the discovery of ET. There is no plausibility whatsoever to scenarios of alien visitations and abductions, though it can make for good science fiction as any viewer of *The X-Files* knows.

Computer science

John Horgan interviewed many visionaries, some of whom look to computers as a tool that will enable us to expand science in ways that are totally unlike those we experience today. As computer science is my official field of training and expertise, I will allow myself to comment upon these visions. We computer scientists have become used to justifying our failings by claiming that we are engaged in a very new science. But that excuse is wearing thin now that over fifty years have passed since the first digital computers were built. Fifty years is about the same length of time as from the discovery of radioactivity through the development of the theory of quantum mechanics to the achievement of controlled and uncontrolled nuclear energy. With a fifty-year perspective, it is clear that the "giant brains" have not lived up to their hype. Computers are excellent tools not only for calculation but also for control of machinery such as telephones and aircraft. They can be used to simulate (with varying degrees of fidelity) physical systems that are too dangerous or too large to investigate directly. Furthermore, they have enabled ordinary people to engage in information processing (word processing, graphics, databases, and communications) that was originally limited to commercial enterprises that could afford to employ specialized technicians using expensive machinery.

But the field of artificial intelligence has barely progressed beyond demonstrations or useful applications that display little, if any, cognitive ability. It is hardly shocking that a computer can successfully play a game like chess that is defined by arbitrary formal rules on a small 8x8 grid. An anecdote may help make this distinction between computation and cognition clear. I once saw a chess match between world champion Gary Kasparov (1963–) and the IBM chess program Deep Blue. At one point, Kasparov, who had removed his watch, strapped it back onto his wrist. The commentator mentioned that Kasparov does this somewhat absentmindedly as a sign of impatience when he knows that a victory for him is inevitable. Deep Blue, of course, could know nothing of the sort, and some minutes went by before its human "keeper" (one of the programmers who developed Deep Blue) decided to capitulate.

Still there is no doubt that software can become even more complex, and this leads to speculation that computers will eventually become more intelligent than we are and that perhaps they will take over the world. I regard this as nonsense. From the same fifty-year perspective, one can sense disappointment not only in the achievements of speculative fields like artificial intelligence, but even in the progress of basic software technology. So-called new languages and systems have their roots in very early advances that were made in the late sixties and early seventies when computer technology was in its infancy. Even the technology of e-mail and information transfer over the Internet is not unrecognizably different from pioneering work done in the early seventies.

But more important is the fact that for software to do anything it must be executed on a computer. Suppose that a cognitively conscious computer tried to take over California. Exactly how would it ensure a continuous supply of electrical power? Would it invade Oregon or Nevada to demand that they supply it? Would it hold a gun to the heads of planning councils until they agreed to build a nuclear power plant? Perhaps it would use solar cells, but then it would be at the mercy of smog and kids armed with blankets and spray paint. We are told that such computers will be able to maintain themselves, by replacing defective circuits. Are we to suppose that a computer will don a hard hat and rubber boots and descend into the Earth to mine copper? Or that it would hire scabs to break a strike at a semiconductor fabrication plant that made its spare chips? Steven Harris suggests just this sort of scenario:

> If a super-intelligent computer has enough contact with the world to be very useful, it will probably have enough contact to corrupt some of its captors into allowing it to escape. An artificial intelligence might amass wealth, for example, and with that wealth influence the passage of laws in democracies. It might also simply bribe outlaw humans and outlaw governments.[7]

I suppose that we are to assume that humans will take this lying down, and not respond with political corruption and bribery of their own, to say nothing of the violence of which we humans are the unequalled masters. The masters of this planet are, and for a long time to come will remain—barring a nuclear war or an asteroid collision—members of our species. Computers are just a tool, albeit one that can amplify our ability to do good or evil, but not a danger nor a hope in and of themselves.

* * *

While many, many surprises probably await scientific discovery, true revolutions will be few, and even those will almost certainly deal with extreme conditions that are unlikely to be interesting to anyone but scientists. Nevertheless, the surprises and even the seemingly mundane advances in science are fascinating, and a young student considering a career in science should not be put off by pessimism, justified or not, about revolutions. You can still win a Nobel Prize and you can still find fulfillment in expanding our understanding of the universe.

EDWARD WITTEN: THE EINSTEIN OF TODAY

Edward Witten (1951–) has been called the most brilliant physicist since Albert Einstein. Like Einstein in his time, Witten works at the Institute for Advanced Study in Princeton, New Jersey, and like Einstein, he is associated with a revolutionary new theory in physics. But here the similarity ends.

Witten only began the study of physics in graduate school, while Einstein always wanted to be a physicist. Einstein pioneered the field of relativity, while Witten came to the field of the theory of strings or superstrings only after it had been invented. He was drawn to string theory, because the existence of gravity is a natural consequence of the theory, which makes it a prime candidate for unifying the two main areas of twentieth-century physics: quantum mechanics and relativity. Since he began working in string theory, Edward Witten has become its most prolific researcher, and his papers are cited thousands of times by scientists in the field.

By the mid-1990s, string theory had become somewhat confused and stagnant, with five competing versions, none of which seemed to have any significant advantage over any of the others. Edward Witten showed that the five theories could all be seen as limiting cases of a unified string theory called *M-theory*; this advance reanimated research into string theory.

Perhaps the most significant difference between Witten and Einstein is that Einstein was first and foremost a physicist, and although competent in mathematics never saw himself as an expert; he frequently asked for the help of professional mathematicians like Hermann Minkowski (1864–1909) and Marcel Grossmann (1878–1936). Witten, on the other hand, won the 1990 Fields Medal, the highest award in mathematics.[8] The combination of physical insight and mathematical talent enabled him to make fundamental contributions to the modern mathematics of topology and geometry.

It is precisely the extreme mathematization of physics inherent in string theory that makes it somewhat controversial. While all scientists since the days of Galileo and Newton understand the importance of the mathematization of science, it is hard to accept a theory of physics whose foremost practitioner is a mathematician of the caliber needed to win a Fields Medal. Sheldon Glashow (1932–), who won a Nobel Prize for Physics in 1979, professes himself unable to understand string theory and very wary of a theory whose predictions are not experimentally testable.[9]

If in the future, string theory eventually becomes accepted as a correct description of nature by the community of physicists, perhaps as the result of a prediction that is experimentally verified, Edward Witten will be able to add a Nobel Prize in Physics to his Fields Medal.

Notes

(Where electronic texts are referenced, no page numbers are given.)

Preface

1. This aphorism was attributed by Carl Sagan to James Oberg, but apparently was not original with him. See James Hrynyshyn, "When Brains Fall Out: The Origin of a Skeptical Maxim," *Skeptic* 10, no. 1 (2003): 34–35.

Chapter 1

1. http://www.randi.org/jr/05-15-2000.html (accessed January 1, 2005).
2. Rose-Marie Hagen, *What Great Paintings Say*, Vol. 2 (New York: Taschen, 1997).
3. Alan F. Chalmers, *What is This Thing Called Science?* 3rd ed. (Buckingham: Open University Press, 1999), pp. 38–39.
4. Image courtesy IBM Corporation. Unauthorized use not permitted. For more images obtained using STMs, see http://www.almaden.ibm.com/vis/stm/gallery.html (accessed January 1, 2005).
5. I will use the term *Newtonian mechanics* for Isaac Newton's theory that includes the law of gravitation and the three laws of motion.

Chapter 2

1. More precisely, the Earth and the Sun revolve around a common center of gravity, the *barycenter*, which is so near the Sun that for all practical purposes we can say that the Earth revolves around the Sun.
2. Holton and Brush prefer to translate "feign no hypotheses" instead of "frame no hypotheses." *Feign* means to assume or suggest something that you don't really believe in. Gerald Holton and Stephen G. Brush, *Physics, the Human Adventure: From Copernicus to Einstein and Beyond* (New Brunswick, NJ: Rutgers University Press, 2001), p. 108.
3. If you do not have the time to read the entire book, the final chapter gives a readable "recapitulation" of his arguments.
4. A comprehensive discussion of the predictive power of evolution can be found in Douglas Theobald, "29+ Evidences for Macroevolution: The Scientific Case for the Theory of Common Descent with Gradual Modification," http://www.talkorigins.org/faqs/comdesc (accessed January 1, 2005).
5. http://www.cdc.gov/malaria/ (accessed January 1, 2005).
6. Charles Darwin, *On the Origin of Species by Means of Natural Selection, or the Preservation of Favoured Races in the Struggle for Life*, 6th ed. (London: John Murray, 1872), chap. 15.
7. For more information on coelacanths, see http://www.pbs.org/wgbh/nova/fish (accessed January 1, 2005), http://www.mnh.si.edu/highlight/coelacanth (accessed January 1, 2005) and their own Web site http://www.dinofish.com (accessed January 1, 2005).
8. Michael Ruse, *The Darwinian Revolution: Science Red in Tooth and Claw*, 2nd ed. (Chicago: University of Chicago Press, 1999), p. 175.
9. See Niles Eldredge, *The Triumph of Evolution and the Failure of Creationism* (New York: Freeman, 2000); Douglas Futuyama, *Science on Trial: The Case for Evolution* (Sunder-

land, MA: Sinauer Associates, 1995); and Theobald, "29+ Evidences for Macroevolution" for a discussion of the evidence for evolution and a refutation of the creationist claims against the theory. Robert T. Pennock, *Tower of Babel: The Evidence Against the New Creationism* (Cambridge, MA: MIT Press, 1999) is a comprehensive analysis of creationism from the point of view of a philosopher.

10. Massimo Pigliucci, "The Case Against God: Science and the Falsifiability Question in Theology," *Skeptic* 6, no. 2 (1998): 66–73.

11. Henry Morris and Gary E. Parker, *What Is Creation Science?* (El Cajon, CA: Master Books, 1987), p. 18.

12. Ibid.

13. Ibid., p. xii.

14. I must admit to some embarrassment when, after writing this paragraph, I found an eminent paleontologist writing of "evolutionists like me" (Eldredge, *The Triumph of Evolution*, p. 162).

15. John Gribbin, "On the Shoulders of Midgets," *Skeptic* 10, no. 1 (2003): 36–39.

Chapter 3

1. For a discussion of the difference between theories and laws within the context of biology, see William F. McComas, "A Textbook Case of the Nature of Science: Laws and Theories in the Science of Biology," *International Journal of Science and Mathematics Education* 1 (2003): 141–55.

2. Carl Sagan, *The Demon Haunted World: Science as a Candle in the Dark* (London: Headline, 1997), p. 78.

3. Gregg Easterbrook, "Science Sees the Light," *New Republic* 219, no. 15 (1998): 24, 26, 27.

4. Brian Greene, *The Elegant Universe: Superstrings, Hidden Dimensions, and the Quest for the Ultimate Theory* (New York: Norton, 1999), p. 5.

5. See Stephen Jay Gould, *Wonderful Life: The Burgess Shale and the Nature of History* (New York: Norton, 1989) for an extensive discussion of the lack of purpose in evolution.

6. See Peter J. Bowler, *Charles Darwin: The Man and his Influence* (Oxford: Blackwell, 1990), p. 171. Bowler's book and Ruse, *The Darwinian Revolution* describe the political, intellectual, and social context at the time of the publication of Darwin's theory.

7. The ship's surgeon was the official naturalist for the voyage.

8. There is an excellent Web site on Darwin: http://www.aboutdarwin.com/ (accessed January 1, 2005).

Chapter 4

1. See chapters 5–7 of Chalmers, *What is This Thing Called Science?*, for an extensive and accessible discussion of falsification.

2. For the opposing view, see Kevin MacDonald, "Psychoanalysis as Pseudoscience," in *The Skeptic Encyclopedia of Pseudoscience*, ed. Michael Shermer (Santa Barbara, CA: ABC-CLIO, 2002), pp. 373–83.

3. Gerald Holton and Stephen G. Brush, *Physics, the Human Adventure: From Copernicus to Einstein and Beyond* (New Brunswick, NJ: Rutgers University Press, 2001), chap. 4.

4. For more details on this episode, see chap. 3 of Gerald Holton, *Science and Anti-Science* (Cambridge, MA: Harvard University Press, 1993).

5. Adapted from Holton and Brush, *Physics, the Human Adventure*, pp. 135–36. The answer is given immediately after the exercise!

6. See A. K. Dewdney, *Yes, We Have No Neutrons: An Eye-Opening Tour Through the Twists and Turns of Bad Science* (New York: Wiley, 1997), chap. 1.

7. Darwin, *On the Origin of Species*, chap. 15.

8. Ibid.

9. Rebecca Stott, *Darwin and the Barnacle: The Story of One Tiny Creature and History's Most Spectacular Scientific Breakthrough* (New York: Norton, 2003).

10. Photos courtesy of Michael Silver. His Web site, http://www.barnacle.com (accessed January 1, 2005), contains many other images.

11. Francis Darwin, ed., *The Autobiography of Charles Darwin: From The Life and Letters of Charles Darwin*, (London: John Murray, 1887).

12. See Theobald, "29+ Evidences for Macroevolution."

13. Henry Morris and Gary E. Parker, *What Is Creation Science?* (El Cajon, CA: Master Books, 1987), p. 224.

14. See Kate Wong, "The Mammals That Conquered the Seas," *Scientific American* 286, no. 5 (2002): 70–79 for details of the transitional sequence and drawings of these animals.

15. Dennis Overbye, *Einstein in Love: A Scientific Romance* (New York: Penguin Books, 2001), p. 7.

16. Ibid., p. 135.

17. Ibid., p. 214.

Chapter 5

1. Raymond J. Seeger, *Galileo Galilei: His Life and His Works* (Oxford: Pergamon Press, 1966), p. 13.

2. Dana Densmore and William Donahue, *Newton's Principia—The Central Argument: Translation, Notes, and Expanded Proofs* (Santa Fe, NM: Green Lion Press, 1995).

3. See Seeger, *Galileo Galilei*, chap. 16.

4. Mark Ridley, *Evolution* (Cambridge, MA: Blackwell Scientific, 1993).

5. http://www.fda.gov/fdac/features/096_home.html (accessed January 1, 2005).

6. Images courtesy of Jim Kaler. See the Web sites http://www.astro.uiuc.edu/~kaler/sow/ sow.html (accessed January 1, 2005) and http://www.astro.wisc.edu/~dolan/constellations/ (accessed January 1, 2005) for more information on constellations.

7. This is nicely explained in Julius Staal, *The New Patterns in the Sky: Myths and Legends of the Stars* (Blacksburg, VA: McDonald & Woodward, 1988).

8. See Roger B. Culver and Philip A. Ianna, *Astrology: True or False?: A Scientific Evaluation* (Amherst, NY: Prometheus Books, 1988) for a summary of these experiments.

9. *A Primer on Medical Device Interactions with Magnetic Resonance Imaging Systems*, Food and Drug Administration, Center for Devices and Radiological Health, http://www.fda.gov/ cdrh/ode/primerf6.html (accessed January 1, 2005). There are also potential dangers from heating caused by the electrical fields in the apparatus, but these are not essentially different from the dangers of sunstroke or dehydration.

10. Benjamin Franklin and Antoine Lavoisier, "Report of the Commissioners Charged by the King to Examine Animal Magnetism, Printed on the King's Order Number 4 in Paris from the Royal Printing House," in Michael Shermer, ed., *The Skeptic Encyclopedia of Pseudo-science* (Santa Barbara, CA: ABC-CLIO, 2002), pp. 797–821.

11. Culver and Ianna, *Astrology*, p. 105. Dean and Kelley report on an interesting study in which over 2,000 people born in the same hospital just a few days apart from each other were examined over twenty years later. They found no correlation between closeness of the time of birth and any physical, mental, or personality traits. Geoffrey Dean and Ivan W. Kelly, "Is Astrology Relevant to Consciousness and Psi?" *Journal of Consciousness Studies* 10, nos. 6–7 (2003): 175–98.

12. See http://www.homeowatch.org (accessed January 1, 2005) for more details.

13. http://vm.cfsan.fda.gov (accessed January 1, 2005).

14. Michael Shermer, "Psychic for a Day: Or How I Learned Tarot Cards, Palm Reading, Astrology, and Mediumship in 24 Hours," *Skeptic* 10, no. 1 (2003): 48–55.

15. David Hume, *An Enquiry concerning Human Understanding* (London: A. Millar, 1758), §10.

16. Charles Mackay, *Memoirs of Extraordinary Popular Delusions* (London: Richard Bentley, 1841), end of vol. 2.

17. René Dubos, *Pasteur and Modern Science* (Garden City, NY: Anchor Books, 1960), p. 22. See http://www.labexplorer.com/louis_pasteur.htm (accessed January 1, 2005) for a good online survey of Pasteur's work.

Chapter 6

1. William Shakespeare, *Macbeth*, 5.5.

2. Thomas S. Kuhn, *The Structure of Scientific Revolutions*, 3rd ed. (Chicago: University of Chicago Press, 1996). Kuhn admitted later that the term is problematical, and changed the terminology somewhat.

3. Ibid., p. 150.

4. Ibid.

5. Ibid., p. 103.

6. An accessible treatment of the ups and downs of Galileo's relations with the Church can be found in Dana Sobel, *Galileo's Daughter: A Historical Memoir of Science, Faith, and Love* (New York: Penguin Books, 2000).

7. Dennis Overbye, *Einstein in Love: A Scientific Romance* (New York: Penguin Books, 2001), p. 353.

8. Michael Ruse, *The Darwinian Revolution: Science Red in Tooth and Claw*, 2nd ed. (Chicago: University of Chicago Press, 1999), p. 262.

9. See A. K. Dewdney, *Yes, We Have No Neutrons: An Eye-Opening Tour Through the Twists and Turns of Bad Science* (New York: Wiley, 1997), chap. 6.

10. David Goodstein, "Whatever Happened to Cold Fusion?" California Institute of Technology, 1994, http://www.its.caltech.edu/~dg/fusion.html (accessed January 1, 2005).

11. Energy Research Advisory Board, "Cold Fusion Research: A Report of the Energy Research Advisory Board to the United States Department of Energy," 1989, http://www.ncas.org/erab (accessed January 1, 2005).

12. David Hume, *An Enquiry concerning Human Understanding* (London: A. Millar, 1758), §10.

13. Clark Kimberling, "Emmy Noether," *American Mathematical Monthly* 79, no. 2 (1972): 143.

14. See Claudia Henrion, *Women in Mathematics: The Addition of Difference* (Bloomington: Indiana University Press, 1997) for a presentation of the obstacles facing women mathematicians. While the formal barriers are coming down, there are often little-noticed informal barriers that make it difficult for women to be fully accepted as scientists and mathematicians.

15. See Eugene Hecht, *Physics: Calculus* (Pacific Grove, CA: Brooks/Cole, 1996), pp. 143, 300, 347, 700.

Chapter 7

1. http://www.pugwash.org (accessed January 1, 2005).

2. http://www.wma.net (accessed January 1, 2005).

3. For a simple introduction to postmodernism, see Christopher Butler, *Postmodernism: A Very Short Introduction* (Oxford: Oxford University Press, 2002).

4. David Bloor, *Knowledge and Social Imagery*, 2nd ed. (Chicago: University of Chicago Press, 1991), p. 7.

5. See Paul Gross and Norman Levitt, *Higher Superstition: The Academic Left and Its Quarrels With Science* (Baltimore: Johns Hopkins University Press, 1998) for a strident attack on postmodernism. Noretta Koertge, ed., *A House Built on Sand: Exposing Postmodernist Myths About Science* (New York: Oxford University Press, 1998) is a diverse set of articles defending science from postmodernist criticism.

6. See Steven B. Harris, "The AIDS Heresies: A Case Study in Skepticism Taken Too Far," *Skeptic* 3, no. 2 (1995): 42–58 for an accessible discussion of the science of AIDS and a refutation of the critics. More technical material is available on the Web site of the National Institute of Allergy and Infectious Diseases of the National Institutes of Health: www.niaid.nih.gov (accessed January 1, 2005).

7. Gina, Kolata, *Flu: The Story of the Great Influenza Pandemic of 1918 and the Search for the Virus that Caused It* (New York: Touchstone, 1999). A *pandemic* is an epidemic that affects a wide geographical area.

8. Harris, "The AIDS Heresies."

9. Benno Mueller-Hill, *Murderous Science: Elimination by Scientific Selection of Jews, Gypsies, and Others: Germany 1933–1945* (Oxford: Oxford University Press, 1988).

10. James R. Newman, ed., *The World of Mathematics* (New York: Simon & Schuster, 1956), p. 373.

11. Ibid.

12. The text of this and all articles mentioned in this section, as well dozens of other articles on Sokal's Hoax, can be found on Alan Sokal's Web site: http://www.physics.nyu.edu/faculty/sokal (accessed January 1, 2005).

13. Alan Sokal, "Transgressing the Boundaries: Toward a Transformative Hermeneutics of Quantum Gravity," *Social Text* (Spring-Summer 1996).

14. Ibid.

15. Ibid., n86.

16. Alan Sokal, "A Physicist Experiments With Cultural Studies," *Lingua Franca* (May/June 1996).

17. Bruce Robbins and Andrew Ross, "Response by *Social Text* Editors Bruce Robbins and Andrew Ross," *Lingua Franca* (July/August 1996).

18. Stanley Fish, "Professor Sokal's Bad Joke," *New York Times*, May 21, 1996.

19. Robbins and Ross, "Response by *Social Text* Editors Bruce Robbins and Andrew Ross."

20. For more on Folkman's research, see the *NOVA* program, "Cancer Warrior," http://www.pbs.org/wgbh/nova/transcripts/2805cancer.html (accessed January 1, 2005).

Chapter 8

1. Stephen Jay Gould, "Nonoverlapping Magisteria," *Skeptical Inquirer* 23, no. 4 (1999): 55–61.

2. Ibid., p. 60.

3. Yishayahu Leibowitz, *Conversations on Science and Values*, (Tel Aviv: MOD Publishing, 1985), pp. 7, 35 (in Hebrew, my translation).

4. A recent book explores various approaches to life without the certainty offered by religious belief: Jennifer Michael Hecht, *Doubt: A History* (New York: HarperSanFrancisco, 2003).

5. This is a parody of the 1995 disclaimer on evolution that the state of Alabama required to be pasted in biology textbooks. For more on antievolution legislation in the United States and analyses of this and similar disclaimers, see the Web site of the National Center for Science Education at http://www.ncseweb.org (accessed January 1, 2005) and the Web site of Kenneth Miller at http://www.millerandlevine.com/km/evol (accessed January 1, 2005).

6. Edsger Dijkstra, "Go To Statement Considered Harmful," *Communications of the ACM* 11, no. 3 (1968): 147–48.

7. Dijkstra, "On the Cruelty of Really Teaching Computer Science," *Communications of the ACM* 32, no. 12 (1989): 1398–1404.

Chapter 9

1. http://www.claymath.org (accessed January 1, 2005).

2. James Gleick, *Chaos: Making a New Science* (New York: Viking, 1987).

3. Ibid., p. 304.

4. http://www.nih.gov/news/pr/apr99/niddk-20.htm (accessed January 1, 2005).

5. lpi.oregonstate.edu/infocenter/paulingrec.html (accessed January 1, 2005).

Chapter 10

1. http://www.cancer.org (accessed January 1, 2005).

2. The same data shows that smoking is tied to a high relative morality risk for other cancers (and smoking has been implicated in many other diseases), but for simplicity we limit the discussion to lung cancer.

3. The Broad Street pump handle has acquired scientific cult status similar to the apple that fell on Newton's head. Newton's achievement in developing the theory of gravitation rests not on a single incident, but rather on protracted scientific and mathematical investigation. Similarly, John Snow's achievement rests not on this incident but on protracted epidemiological investigations using field work and statistics. A comprehensive examination of Snow's

life and work has recently been published: Peter Vinten-Johansen, et al., *Cholera, Chloroform, and the Science of Medicine: A Life of John Snow* (New York: Oxford University Press, 2003). It includes a day-by-day chronicle of the pump incident.
4. Photos courtesy of Hezi Ben-Ari.

Chapter 11

1. Raymond M. Smullyan, *What is the Name of This Book? The Riddle of Dracula and Other Logical Puzzles* (Englewood Cliffs, NJ: Prentice-Hall, 1978), p. 184.
2. See chap. 12 of Carl Sagan, *The Demon Haunted World: Science as a Candle in the Dark* (London: Headline, 1997), and the Web sites:
http://www.infidels.org/news/atheism/logic.html (accessed January 1, 2005),
http://www.nizkor.org/features/fallacies (accessed January 1, 2005),
http://www.intrepidsoftware.com/fallacy/welcome.htm (accessed January 1, 2005).
3. Gerald Holton and Stephen G. Brush, *Physics, the Human Adventure: From Copernicus to Einstein and Beyond* (New Brunswick, NJ: Rutgers University Press, 2001), pp. 277–78.
4. Susan Haack, *Defending Science—Within Reason: Between Scientism and Cynicism* (Amherst, NY: Prometheus Books, 2003), pp. 27–28. Haack's thesis of *critical commonsensism* is similar to the position I am taking in this book.
5. Henry Morris and Gary E. Parker, *What Is Creation Science?* (El Cajon, CA: Master Books, 1987), p. 190.
6. http://www.rael.org (accessed January 1, 2005).
7. For a presentation of Gödel's theorems for nonmathematicians, see Smullyan, *What is the Name of This Book?*.
8. Raymond M. Smullyan, *Gödel's Incompleteness Theorems* (New York: Oxford University Press, 1992), p. 109.
9. Even a supergenius like Paul Erdős spent countless sleepless nights trying to solve mathematical problems. He defined a mathematician as "a machine that transforms coffee into theorems." Paul Hoffman, *The Man Who Loved Only Numbers: The Story of Paul Erdős and the Search for Mathematical Truth* (New York: Hyperion, 1998), p. 7.
10. Quoted in Morris Kline, *Mathematics and the Search for Knowledge* (New York: Oxford University Press, 1985), p. 144. Chapters 11–13 of this book discuss the relationship between mathematics and science in detail.
11. Quoted in Simon Singh, *Fermat's Last Theorem* (London: Fourth Estate, 1997), p. 66.
12. The Fields Medal is awarded by the International Mathematical Union. In the absence of a Nobel Prize for mathematics, the Fields Medal is considered to be the highest prize granted in mathematics. While the Nobel Prize is frequently granted to scientists decades after their prize-winning work, the Fields Medal has an age limit in order to encourage the work of young mathematicians. See http://www.mathunion.org/medals (accessed January 1, 2005).
13. Singh, *Fermat's Last Theorem*.
14. Ibid., p. 227.

Chapter 12

1. The Earth is not a true sphere, but rather a shape called an *oblate spheroid*, in which the diameter at the equator is slightly larger than the diameter at the poles. As you go from the poles to the equator, the surface of the Earth rotates faster and faster, and a force is needed

to keep it from flying apart. See Ronald L. Reese, *University Physics* (Pacific Grove, CA: Brooks/Cole, 2000), p. 444–45 for a technical explanation.

2. Quoted by Walter Sullivan, *Continents in Motion: The New Earth Debate*, 2nd ed. (New York: American Institute of Physics, 1991), p. 18.

Chapter 13

1. Quoted by John Horgan, *The End of Science* (London: Abacus, 1996), p. 19.

2. John Maddox, *What Remains to Be Discovered* (London: Papermac, 1998), p. xii.

3. Brian Greene, *The Elegant Universe: Superstrings, Hidden Dimensions, and the Quest for the Ultimate Theory* (New York: Norton, 1999) is an introduction to this theory for the general public. See also the *NOVA* Web site *The Elegant Universe*, http://www.pbs. org/wgbh/nova/elegant/ (accessed January 1, 2005). A competing theory is called *loop quantum gravity*; see Lee Smolin, "Atoms of Space and Time," *Scientific American* 290, no. 1 (2004): 66–75.

4. Image courtesy of Andrew J. Hanson, Indiana University.

5. Maddox, *What Remains to Be Discovered*, p. 41.

6. A lead purveyor of nanotechnology hype is Eric Drexler, *Engines of Creation: The Coming Era of Nanotechnology* (New York: Anchor Books, 1990). See the September 2001 issue of *Scientific American* for a survey of nanotechnology that emphasizes the real scientific achievements and plays down the hype.

7. Steven B. Harris, "A. I. and the Return of the Krell Machine," *Skeptic* 9, no. 3 (2002): 75.

8. For more on the Fields Medal, see the note in the biographical vignette of Andrew Wiles.

9. See http://www.pbs.org/wgbh/nova/elegant/view-glashow.html (accessed January 1, 2005).

Recommended Reading

Blackburn, Simon. *Think: A Compelling Introduction to Philosophy.* New York: Oxford University Press, 1999. This is an elementary introduction to philosophy.

Chalmers, Alan F. *What is This Thing Called Science?* 3rd ed. Buckingham: Open University Press, 1999. This excellent introduction to the philosophy of science is my recommendation for the next book to read on the nature of science.

Culver, Roger B., and Philip A. Ianna. *Astrology: True or False?; A Scientific Evaluation.* Amherst, NY: Prometheus Books, 1988. Culver and Ianna are professional astronomers who thoroughly debunk astrology.

Gross, Paul R., and Norman Levitt. *Higher Superstition: The Academic Left and Its Quarrels With Science.* Baltimore: Johns Hopkins University Press, 1998. A passionate refutation of postmodernist criticism of science.

Holton, Gerald, and Stephen G. Brush. *Physics, the Human Adventure: From Copernicus to Einstein and Beyond.* New Brunswick, NJ: Rutgers University Press, 2001. This textbook presents a comprehensive overview of physical science from a philosophical and historical perspective.

Hume, David. *Dialogues Concerning Natural Religion.* London: Robinson, 1779. Available from Gutenberg at http://www.gutenberg.net (accessed January 1, 2005). Over two centuries ago, Hume provided coherent arguments for a rational approach to science. He demolished the argument from design decades before Darwin even thought about evolution.

Kline, Morris. *Mathematics: The Loss of Certainty.* New York: Oxford University Press, 1980. Kline portrays the history of mathematics up to the rise of rigor in the nineteenth century and the twentieth-century crises in its foundations.

Mackay, Charles. *Memoirs of Extraordinary Popular Delusions.* London: Richard Bentley, 1841. Available from Gutenberg at http://www.gutenberg.net (accessed January 1, 2005). Pseudoscience and attacks on pseudoscience are not new. If you are not deterred by the long-winded, florid nineteenth-century prose, this book is quite charming.

Matthews, Michael R. *Science Teaching: The Role of History and Philosophy of Science.* New York: Routledge, 1994. This book discusses the history and philosophy of science from the perspective of a science educator.

National Academy of Sciences. "Teaching About Evolution and the Nature of Science." 1998. http://www.nap.edu/books/0309063647/html/index.html (accessed January 1, 2005). The NAS presents evolution from the teacher's point of view.

Okasha, Samir. *Philosophy of Science: A Very Short Introduction.* Oxford: Oxford University Press, 2002. An introduction to the philosophy of science that is more elementary than the Chalmers book.

Pennock, Robert T. *Tower of Babel: The Evidence Against the New Creationism.* Cambridge, MA: MIT Press, 1999. Pennock analyzes the creationist arguments in detail; it is a must-read for people who need to debate creationists.

Pigliucci, Massimo. "Personal Gods, Deism, and the Limits of Skepticism." *Skeptic* 8, no. 2 (2000): 38–45. An excellent survey of various positions on the relationship between religion and science.

Plait, Philip. *Bad Astronomy: Misconceptions And Misuses Revealed, From Astrology To The Moon Landing "Hoax!"* New York: Wiley, 2002. Plait clears up common misconceptions on astronomy and provides a detailed debunking of the claim that the Apollo moon landings were faked.

Randi, James. *Flim Flam: The Truth About Parapsychology, Unicorns, and Other Delusions.* Amherst, NY: Prometheus Books, 1980. James Randi he tells of his battles with pseudoscientists.

Sagan, Carl. *The Demon Haunted World: Science as a Candle in the Dark.* London: Headline, 1997. Sagan, famous for his writings and films on science, takes on the pseudoscientists, especially UFOlogists.

Shermer, Michael. *Why People Believe Weird Things : Pseudoscience, Superstition, and Other Confusions of Our Time.* New York: Freeman, 1997. An excellent analysis of the attraction of pseudoscience.

———, ed. *The Skeptic Encyclopedia of Pseudoscience.* Santa Barbara, CA: ABC-CLIO, 2002. This source book contains articles on the various pseudosciences written by experts in each field. It also includes original documents, and debates on topics whose scientific status is in doubt.

Smullyan, Raymond M. *What is the Name of This Book? The Riddle of Dracula and Other Logical Puzzles.* Englewood Cliffs, NJ: Prentice-Hall, 1978. A delightful book of logical puzzles that can help you practice clear thinking.

Snow, John. *On the Mode of Communication of Cholera.* 2nd ed. London: Churchill, 1855. Available at http://www.ph.ucla.edu/epi/snow.html (accessed January 1, 2005). Snow's book on his war against cholera is fascinating to read even 150 years later.

Stott, Rebecca. *Darwin and the Barnacle: The Story of One Tiny Creature and History's Most Spectacular Scientific Breakthrough.* New York: Norton, 2003. This is a detailed chronicle of the life and work of Charles Darwin during the eight years that he studied barnacles in depth before the publication of the *Origin.*

US Geological Survey. "The Dynamic Earth: The Story of Plate Tectonics." 2001. pubs.usgs.gov/publications/text/dynamic.html (accessed January 1, 2005). The USGS has provided a short, colorful introduction to continental drift and plate tectonics.

Bibliography

Bloor, David. *Knowledge and Social Imagery.* 2nd ed. Chicago: University of Chicago Press, 1991.

Bowler, Peter J. *Charles Darwin: The Man and His Influence.* Oxford: Blackwell, 1990.

Butler, Christopher. *Postmodernism: A Very Short Introduction.* Oxford: Oxford University Press, 2002.

Christianson, Gale E. *Isaac Newton and the Scientific Revolution.* Oxford Portraits in Science. New York: Oxford University Press, 1996.

Darwin, Charles. *On the Origin of Species by Means of Natural Selection, or the Preservation of Favoured Races in the Struggle for Life.* 6th ed. London: John Murray, 1872. Available from Gutenberg at http://www.gutenberg.net (accessed January 1, 2005).

Darwin, Francis, ed. *The Autobiography of Charles Darwin: From "The Life and Letters of Charles Darwin."* London: John Murray, 1887. Available from Gutenberg at http://www.gutenberg.net (accessed January 1, 2005).

Dean, Geoffrey, and Ivan W. Kelly. "Is Astrology Relevant to Consciousness and Psi?" *Journal of Consciousness Studies* 10, nos. 6–7 (2003): 175–198.

Densmore, Dana, and William Donahue. *Newton's Principia—The Central Argument: Translation, Notes, and Expanded Proofs.* Santa Fe, NM: Green Lion Press, 1995.

Dewdney, A. K. *Yes, We Have No Neutrons: An Eye-Opening Tour Through the Twists and Turns of Bad Science.* New York: Wiley, 1997.

Dijkstra, Edsger. "Go To Statement Considered Harmful." *Communications of the ACM* 11, no. 3 (1968): 147–48.

———. ."On the Cruelty of Really Teaching Computer Science." *Communications of the ACM* 32, no. 12 (1989): 1398–1404.

Drexler, K. Eric. *Engines of Creation: The Coming Era of Nanotechnology.* New York: Anchor Books, 1990.

Dubos, René. *Pasteur and Modern Science.* Garden City, NY: Anchor Books, 1960.

Easterbrook, Gregg. "Science Sees the Light." *New Republic* 219, no. 15 (1998): 24–29.

Eldredge, Niles. *The Triumph of Evolution and the Failure of Creationism.* New York: Freeman, 2000.

Energy Research Advisory Board. "Cold Fusion Research: A Report of the Energy Research Advisory Board to the United States Department of Energy." 1989. http://www.ncas.org/erab (accessed January 1, 2005).

Feyerabend, Paul. *Against Method.* London: Verso, 1993.

Fish, Stanley. "Professor Sokal's Bad Joke." *New York Times*, May 21, 1996.

Futuyama, Douglas. *Science on Trial: The Case for Evolution*. Sunderland, MA: Sinauer Associates, 1995.

Gleick, James. *Chaos: Making a New Science*. New York: Viking, 1987.

Goodstein, David. "Whatever Happened to Cold Fusion?" California Institute of Technology. 1994. http://www.its.caltech.edu/~dg/fusion.html (accessed January 1, 2005).

Gould, Stephen Jay. "Nonoverlapping Magisteria." *Natural History* 106, no. 2 (1997): 16–22. Also in: *Skeptical Inquirer* 23, no. 4 (1999): 55–61.

———. *Wonderful Life: The Burgess Shale and the Nature of History*. New York: Norton, 1989.

Greene, Brian. *The Elegant Universe: Superstrings, Hidden Dimensions, and the Quest for the Ultimate Theory*. New York: Norton, 1999.

Gribbin, John. "On the Shoulders of Midgets." *Skeptic* 10, no. 1 (2003): 36–39.

Haack, Susan. *Defending Science—Within Reason: Between Scientism and Cynicism*. Amherst, NY: Prometheus Books, 2003.

Hagen, Rose-Marie. *What Great Paintings Say*. Vol. 2. New York: Taschen, 1997.

Harris, Steven B. "The AIDS Heresies: A Case Study in Skepticism Taken Too Far." *Skeptic* 3, no. 2 (1995): 42–58.

———. "A. I. and the Return of the Krell Machine." *Skeptic* 9, no. 3 (2002): 68–79.

Hecht, Eugene. *Physics: Calculus*. Pacific Grove, CA: Brooks/Cole, 1996.

Hecht, Jennifer Michael. *Doubt: A History; The Great Doubters and Their Legacy of Innovation from Socrates and Jesus to Thomas Jefferson and Emily Dickinson*. New York: HarperSanFrancisco, 2003.

Henrion, Claudia. *Women in Mathematics: The Addition of Difference*. Bloomington, IN: Indiana University Press, 1997.

Hoffman, Paul. *The Man Who Loved Only Numbers: The Story of Paul Erdős and the Search for Mathematical Truth*. New York: Hyperion, 1998.

Holton, Gerald. *Science and Anti-Science*. Cambridge, MA: Harvard University Press, 1993.

Horgan, John. *The End of Science*. London: Abacus, 1996.

Hume, David. *An Enquiry concerning Human Understanding*. London: A. Millar, 1758. Available at http://www.ecn.bris.ac.uk/het/hume/enquiry (accessed January 1, 2005).

Hrynyshyn, James. "When Brains Fall Out: The Origin of a Skeptical Maxim." *Skeptic* 10, no. 1 (2003): 34–35.

Isaac, Richard. *The Pleasures of Probability*. New York: Springer-Verlag, 1995.

Kanigel, Robert. *The Man Who Knew Infinity: A Life of the Genius Ramanujan*. New York: Washington Square Press, 1991.

Kimberling, Clark. "Emmy Noether." *American Mathematical Monthly* 79, no. 2 (1972): 136–49.

Kline, Morris. *Mathematics and the Search for Knowledge.* New York: Oxford University Press, 1985.

Koertge, Noretta, ed. *A House Built on Sand: Exposing Postmodernist Myths About Science.* New York: Oxford University Press, 1998.

Kolata, Gina. *Flu: The Story of the Great Influenza Pandemic of 1918 and the Search for the Virus that Caused It.* New York: Touchstone, 1999.

Kuhn, Thomas S. *The Structure of Scientific Revolutions.* 3rd ed. Chicago: University of Chicago Press, 1996.

Leibowitz, Yishayahu. *Conversations on Science and Values.* Tel Aviv: MOD Publishing, 1985 (in Hebrew).

Maddox, John. *What Remains to Be Discovered.* London: Papermac, 1998.

Matthews, Michael R., ed. *The Scientific Background to Modern Philosophy.* Indianapolis: Hackett, 1989.

McComas, William F. "A Textbook Case of the Nature of Science: Laws and Theories in the Science of Biology." *International Journal of Science and Mathematics Education* 1 (2003): 141–55.

Miller, Kenneth R. *Finding Darwin's God: A Scientist's Search for Common Ground between God and Evolution.* New York: HarperTrade, 1999.

Morris, Henry M. and Gary E. Parker. *What Is Creation Science?* El Cajon, CA: Master Books, 1987.

Mueller-Hill, Benno. *Murderous Science: Elimination by Scientific Selection of Jews, Gypsies, and Others: Germany 1933–1945.* Oxford: Oxford University Press, 1988.

Newman, James R., ed. *The World of Mathematics.* New York: Simon & Schuster, 1956.

Overbye, Dennis. *Einstein in Love: A Scientific Romance.* New York: Penguin Books, 2001.

Pigliucci, Massimo. "The Case Against God: Science and the Falsifiability Question in Theology." *Skeptic* 6, no. 2 (1998): 66–73.

Reese, Ronald L. *University Physics.* Pacific Grove, CA: Brooks/Cole, 2000.

Ridley, Mark. *Evolution.* Cambridge, MA: Blackwell Scientific, 1993.

Robbins, Bruce, and Andrew Ross. "Response by *Social Text* editors Bruce Robbins and Andrew Ross." *Lingua Franca* (July/August 1996).

Ruse, Michael. *The Darwinian Revolution: Science Red in Tooth and Claw.* 2nd ed. Chicago: University of Chicago Press, 1999.

Seeger, Raymond J. *Galileo Galilei: His Life and His Works.* Oxford: Pergamon Press, 1966.

Shasha, Dennis, and Cathy Lazere. *Out of Their Minds: The Lives and Discoveries of 15 Great Computer Scientists.* New York: Copernicus, 1995.

Shermer, Michael. "Psychic for a Day: Or How I Learned Tarot Cards, Palm Reading, Astrology, and Mediumship in 24 Hours." *Skeptic* 10, no. 1 (2003): 48–55.

Singh, Simon. *Fermat's Last Theorem*. London: Fourth Estate, 1997.

Smolin, Lee. "Atoms of Space and Time." *Scientific American* 290, no. 1 (2004): 66–75.

Smullyan, Raymond M. *Gödel's Incompleteness Theorems*. New York: Oxford University Press, 1992.

Sobel, Dana. *Galileo's Daughter: A Historical Memoir of Science, Faith, and Love*. New York: Penguin Books, 2000.

Sokal, Alan. "A Physicist Experiments With Cultural Studies." *Lingua Franca* (May/June 1996).

———. "Transgressing the Boundaries: Toward a Transformative Hermeneutics of Quantum Gravity." *Social Text* (Spring-Summer 1996).

Staal, Julius. *The New Patterns in the Sky: Myths and Legends of the Stars*. Blacksburg, VA: McDonald & Woodward, 1988.

Sullivan, Walter. *Continents in Motion: The New Earth Debate*. 2nd ed. New York: American Institute of Physics, 1991.

Theobald, Douglas. "29+ Evidences for Macroevolution: The Scientific Case for the Theory of Common Descent with Gradual Modification." 2002. http://www.talkorigins.org/faqs/comdesc (accessed January 1, 2005).

Vinten-Johansen, Peter, Howard Brody, Nigel Paneth, Stephen Rachman, and Michael Rip. *Cholera, Chloroform, and the Science of Medicine: A Life of John Snow*. New York: Oxford University Press, 2003.

Wong, Kate. "The Mammals That Conquered the Seas." *Scientific American* 286, no. 5 (2002): 70–79.

Index